IET ENERGY ENGINEERING SERIES 225

Fusion–Fission Hybrid Nuclear Reactors

Weston M. Stacey

Emeritus Regent's Professor of Nuclear Engineering
Georgia Institute of Technology

Fusion–Fission Hybrid Nuclear Reactors

For enhanced nuclear fuel utilization and radioactive waste reduction

Weston M. Stacey

Emeritus Regent's Professor of Nuclear Engineering
Georgia Institute of Technology

The Institution of Engineering and Technology

Published by The Institution of Engineering and Technology, London, United Kingdom

The Institution of Engineering and Technology is registered as a Charity in England & Wales (no. 211014) and Scotland (no. SC038698).

The Institution of Engineering and Technology
Futures Place
Kings Way, Stevenage
Hertfordshire, SG1 2UA, United Kingdom

www.theiet.org

British Library Cataloguing in Publication Data
A catalogue record for this product is available from the British Library

ISBN 978-1-83953-651-9 (hardback)
ISBN 978-1-83953-652-6 (PDF)

Typeset in India by MPS

Cover photo is of the European JET Tokamak at Culham in the United Kingdom.

Contents

List of figures

List of tables

About the author

Weston M. (Bil) Stacey is emeritus Regent's Professor of Nuclear Engineering at Georgia Institute of Technology, United States. His career in nuclear science and engineering spans 60+ years of research in nuclear reactor physics and nuclear reactor design, fusion plasma physics and fusion reactor design, at Knolls Atomic Power Laboratory, Argonne National Laboratory and Georgia Tech. He led the international IAEA INTOR Workshop (USA, USSR, EU, Japan, 1977–88) project to assess the readiness of the world's fusion programs to build an experimental fusion power reactor, to evaluate the required additional R&D, and to develop the INTOR conceptual design; this project evolved into the current ITER project for the first experimental fusion reactor. He is author of more than 350 research papers and 7 books on nuclear fission and fusion physics and technology and recipient of the ANS Seaborg and Wigner Reactor Physics awards, the DOE Distinguished Associate award and two Certificates of Appreciation, the Fusion Power Assoc. Distinguished Career award, and is a lifetime member of Who's Who.

Acknowledgments

The subject of this book, the fusion–fission hybrid nuclear reactor, has great potential for significantly enhancing our use of clean nuclear energy resources, and was the subject of several MS and PhD theses and also several student–faculty design projects at Georgia Tech over the past two decades, which have been instrumental in the development of much of the material presented in this book. The author would like to acknowledge in particular the direct contributions and advice of *Professors Bojan Petrovic and Wilfred Van Rooijen* in the areas of nuclear reactor physics and fuel cycle design; the advice of *Professors S. M. Ghiassian and A. S. Erickson* on reactor heat removal; and the important research contributions of former graduate students *Drs. Edward Hoffman, Andrew Bopp, John-Patrick Floyd, Chris Sommer, Chris Stewart, Tyler Sumner, Theresa Wilks, Alex Moore and Mr. James Maddox*. The contributions of many colleagues world-wide are attested in the references, and the summary of previous magnetic mirror work by Dr. Ralph Moir is particularly noted. Finally, there would not be a book without the invaluable assistance of *Mr. Will DeShazer* in the extraction from the literature and physical assembly of the various figures, tables and equations used in a manuscript that could be shared electronically amongst us and with the publisher. Finally, the efforts and helpful suggestions of *Christoph von Friedeburg* (Books Commissioning Editor, IET), *Olivia Wilkins* (Assistant Editor, IET), Natalie Harper (Production Assistant, IET), and *Srinivasan N.* (Project Manager, MPS Ltd) have played essential roles in bringing this book into existence in its final form, all of which are gratefully acknowledged.

Preface

This book provides a broad technical examination and evaluation of combining nuclear fission and nuclear fusion sciences and technologies to create enhanced fusion–fission hybrid nuclear reactors for electric power production, the destruction of "nuclear waste" and the breeding of nuclear fissionable fuel.

The perceived technical benefit of combining copious fusion neutron sources and neutron-sparse subcritical nuclear fission reactors (into what is called a "fusion– fission hybrid" (FFH) nuclear reactor) is to have more neutrons to work with in the nuclear fission reactor. These "extra" neutrons from fusion can be used to (1) "breed" fissionable plutonium and uranium isotopes from essentially non-fissionable U238 and totally non-fissionable Th232, which constitute the over-whelming majority of uranium and all of thorium ores, respectively. These bred fissionable isotopes can then be neutron-fissioned, in the same or other nuclear fission reactors, thereby substantially increasing (by a factor of maybe 10–50, depending on the fuel cycle and other technical factors) the nuclear fission energy recoverable from a given amount of uranium ore and enabling the recovery of the potential nuclear energy from the otherwise non-fissionable thorium ore. This new, neutron-rich fission reactor "breeder" fuel cycle enabled by the fusion neutron source in FFH would increase from several decades to several millennia the time that the known nuclear fission energy resources could provide the world's present total electricity production.

The second use of these "extra fusion neutrons," to fission (hence both obtain additional energy from and destroy) the highly radioactive and extremely long-lived radioactive actinide elements remaining in "spent" nuclear fuel, is not widely recognized. These long-lived radioactive actinide elements constitute most of the so-called "high-level nuclear waste". Thereby fissioning them not only extracts extra energy from that "not-quite spent" fuel but also destroys much of the highly radio-active material that would otherwise need to be stored for millennia in long-term "high-level" waste repositories (HLWRs; e.g., Yucca Mountain). Such "burner" FFH reactors would dramatically reduce the world's "high-level" (nuclear) waste reposi-tory requirements, while producing additional nuclear energy.

Hence, the three major anticipated benefits of the FFH subcritical fission reactor concept are (1) enabling much more efficient utilization of the nuclear energy content of the world's known uranium ore; in fact, extending the capability of that ore to produce the present total world electricity production from several decades to several millennia; (2) gaining access to the comparable potential nuclear energy content of the world's thorium ore; and (3) substantially reducing the amount of high-level waste that must be buried in HLWRs like Yucca Mountain, in the process significantly increasing the energy obtained from the uranium and thorium ores.

There is also a practical benefit of the FFH to the world's developing fusion power program. Notwithstanding the optimistic projections by many promoters of "superior" new fusion plasma and fusion technology concepts "just around the corner," experience with other such concepts over the years suggests that it is going to take many years and a lot of money to further develop fusion plasma physics, materials, magnet technologies, tritium recovery technologies, etc., to the point where fusion can produce *economical and reliable* electrical power on its own.

However, essentially the same set of plasma physics and fusion technologies are needed for fusion reactors and for the fusion neutron source part of FFH reactors. The substantial economic benefit of fusion neutrons to fission power production through the FFH could be realized beginning in the second half of this century, so that the development of fusion power could, at least in part, "pay its own way" through the FFH.

Detailed design concepts for the SABR FFH reactor concepts, based on IFR/PRISM metal fuel, sodium pool fast reactor technology, and the ITER fusion neutron source physics and technology, are described and employed in this book to carry out fuel cycle, dynamic safety, and other performance analyses of FFH reactors. These analyses support the potential of FFH nuclear reactors to substantially increase the amount of nuclear energy that can be extracted from the uranium and thorium nuclear fuel resources, provide an estimate of the substantial reduction of spent nuclear fuel that must be sent to high-level waste repositories (HLWRs), and illustrate the significant safety advantages of operating nuclear reactors subcritical with a neutron source that can be quickly turned off with an electrical switch to shut down the reactor.

This book is at a level that should be readily comprehensible to anyone with the background of a university upperclassman in applied physics or nuclear, mechanical, electrical, chemical, aeronautical, etc., engineering, as well as (of course) to graduate students and practicing professionals in these and related fields.

The book is also intended to be accessible in the most part to the non-technical reader with an interest in our energy future and realistic options for combatting global warming without disfiguring the surface of the earth. It provides the necessary background material on fission and fusion energy production to make it accessible to the non-technical, as well as, the technical reader. Introduction to both nuclear fission and nuclear fusion power production is included, the mathematics is modest and limited, and the technical discussions are supplemented by informative illustrations.

Other volumes in this series:

Chapter 1

Introduction

A fusion–fission hybrid (FFH) reactor is basically a copious fusion neutron source combined with a subcritical nuclear fission reactor application for those neutrons.

The most commonly mentioned application is a *fusion–fission breeder reactor*, which would capture the fusion neutrons in non-fissionable "fertile" material (U238, which constitutes 99+% of uranium ore) or (Th232, which constitutes 100% of thorium ore) to "breed" (neutron transmute the fertile material into) fissionable material (Pu239 and Pu241 and other fissionable so-called "minor actinides" in the case of U238) or (U233 and U235 and other minor actinides in the case of Th232), for the purpose of subsequently neutron-fissioning these newly "bred" (created) fissionable atoms to produce nuclear fission energy (in the same or a different nuclear fission reactor).

It turns out that the major long-lived radioactive isotopes in "spent" nuclear fuel (the "high-level nuclear waste") are predominantly fissionable actinides (plutonium, americium, etc.), suggesting that a fusion neutron source could be placed next to this "spent" nuclear fuel to reduce the high-level radioactivity of the spent fuel by fissioning it, obtaining about 33% additional nuclear energy from that fuel in the process of reducing the high-level nuclear waste (in a *fusion–fission transmutation or "burner" reactor*).

An appropriate logo for the FFH might well be "more nuclear energy, less nuclear waste."

There are other "nuclear fission" applications of fusion neutrons, hence of FFH reactors, as may be inferred from the titles of the following papers published by faculty and student researchers at the Georgia Institute of Technology together with their national and international colleagues over the last couple of decades. This book draws in large part upon material developed in these papers and similar research papers published by others worldwide, as well as upon relevant textbooks, data references, etc.

Recent Georgia Tech Fusion–Fission Hybrid "FFH" Papers

1. "A Transmutation Facility for Weapons-Grade Plutonium Disposition Based on a Tokamak Fusion Neutron Source"
2. "A Tokamak Tritium Production Reactor"
3. "A Tokamak Tritium Production Reactor Design II"
4. "Capabilities of a DT Tokamak Fusion Neutron Source for Driving a Spent Nuclear Fuel Transmutation Reactor"

5. "A Fusion Transmutation of Waste Reactor"
6. "Nuclear and Fuel Cycle Analysis for a Fusion Transmutation of Waste Reactor"
7. "A Fusion Transmutation of Waste Reactor"
8. "Comparative Fuel Cycle Analysis of Critical and Subcritical Fast Reactor Transmutation Systems"
9. "Nuclear Design and Analysis of the Fusion Transmutation of Waste Reactor"
10. "A Superconducting Tokamak Fusion Transmutation of Waste Reactor"
11. "Nuclear Design and Analysis of the Fusion Transmutation of Waste Reactor"
12. "Subcritical Transmutation Reactors with Tokamak Fusion Neutron Sources"
13. "A Subcritical, Gas-Cooled Fast Transmutation Reactor with a Fusion Neutron Source"
14. "A Subcritical, Helium-Cooled Fast Reactor for the Transmutation of Spent Nuclear Fuel"
15. "Transmutation Missions for Fusion Neutron Sources"
16. "Advances in the Subcritical, Gas-Cooled, Fast Transmutation Reactor Concept"
17. "Tokamak D-T Fusion Neutron Source Requirements for Closing the Nuclear Fuel Cycle"
18. "Fuel Cycle Analysis of a Subcritical Fast Helium-Cooled Transmutation Reactor with a Fusion Neutron Source"
19. "Advances in the Subcritical, Gas-Cooled, Fast Transmutation Reactor Concept"
20. "Sub-critical Transmutation Reactors with Tokamak Fusion Neutron Sources Based on ITER Physics and Technology"
21. "Tokamak Fusion Neutron Source for a Fast Transmutation Reactor"
22. "A TRU-Zr Metal-Fuel Sodium-Cooled Fast Subcritical Advanced Burner Reactor"
23. "Georgia Tech Studies of Sub-critical Advanced Burner Reactors with a D-T Fusion Tokamak Neutron Source for the Transmutation of Spent Fuel"
24. "Dynamic Safety Analysis of the SABR Subcritical Transmutation Reactor Concept"
25. "Tutorial: Principles and Rationale of the Fusion–Fission Hybrid Burner Reactor"
26. "Principles and Rationale of the Fusion–Fission Hybrid Burner Reactor"
27. "Advanced Fuel Cycle Scenario Study in the European Context Using Different Burner Reactors"
28. "Transmutation Fuel Cycle Analyses of the SABR Fission–Fusion Hybrid Burner Reactor for Transuranic and Minor Actinide Fuels"
29. "Resolution of Fission and Fusion Technology Integration Issues: An Updated Design Concept"
30. "The SABrR Concept for a Fission–Fusion Hybrid Fissile Production Reactor"

31. "Solving the Spent Nuclear Fuel Problem by Fissioning Transuranics in Subcritical Advance Burner Reactors"
32. "Dynamic Safety Analysis of a Subcritical Advanced Burner Reactor"
33. "Georgia Tech Studies of Sub-Critical Advanced Burner Reactors with a D-T Fusion Tokamak Neutron Source for the Transmutation of Spent Nuclear Fuel"
    (Complete References provided in Appendix.)

Safety advantages of FFH reactors
Dynamic analyses indicate an unanticipated, but readily understandable, safety benefit of FFHs, which can be turned off within seconds by simply flipping an electric power switch to shut down the fusion neutron source. This simple shut-down mechanism could provide a significant intrinsic safety advantage relative to the normal nuclear fission reactor, which is turned off by mechanically driving neutron-absorbing control rods into the reactor, which is somewhat slower and which, although shown in practice to be very reliable, could conceivably encounter mechanical problems.

Moreover, a subcritical reactor has a much larger reactivity safety margin to prompt super-critical power run-away conditions (the delta-k subcritical) than does a critical reactor (the much smaller delayed neutron fraction, β), which means that the FFHs intrinsically have a larger margin of safety than ordinary critical nuclear reactors against a highly improbable (but not impossible) prompt-supercritical power excursion. A discussion of such differences between critical fission reactors and subcritical FFH reactors based upon above papers [24] and [32] above is included in this book. (I mention this newly recognized and enhanced safety result up-front because it is quite important and not widely recognized, even by most nuclear reactor specialists.)

Chapter 2
Nuclear electric power production

The annual world electric power production (in 2017) was 2.14×10^{13} kWh(e) (214 followed by 11 zeros; kWh(e) is kilowatt-hour electric) and is growing (doubling over the last 20 years) [1]. Most of this recent new power was provided by burning additional carbon-based fuels, unfortunately for mankind because of the associated atmospheric pollution leading to increased global warming with ultimately disastrous consequences, unless it is curtailed very soon.

In the view of most people who have reviewed and understood the facts involved (e.g., National Academy of Sciences committees [2]) there is general agreement *that in order to avoid further calamitous climate change we must stop dumping carbon (which forms carbon dioxide and other "greenhouse" gases that trap escaping reflected solar heat) in the atmosphere, as soon as possible.* This requires the massive worldwide displacement of the burning of carbon-based fuels (coal, oil, gas) to produce thermal and electrical energy with another energy source. There is also general agreement among technically knowledgeable people that, today, nuclear fission energy is the only technically credible and available alternative "clean" energy source that could reliably provide this energy on the scale required at the times and places that it is required, without massive disfiguration of the surface of the earth (hence negative environmental impact of a different type) associated with "harvesting" this "renewable" energy when and where it is available, storing it until it is needed, and then transmitting it to where it is needed.

A major purpose of this book is to document the case that the fuel resources for clean, carbon-free nuclear energy are adequate to provide the world's electrical power for the remainder of this century and, with further technical development of magnetic fusion, for millennia into the future. Both presently available nuclear fission power and future nuclear fusion power, and in particular the combination of the two in subcritical nuclear fission reactors with fusion neutron sources (known as "fusion–fission hybrid" (FFH) reactors), are discussed in detail in this book.

There are presently 448 nuclear fission power reactors worldwide either operating or shortly to be operating, to produce 397,680 MWe (MWe = megawatt electrical = 1 million watts electrical) of electrical power (about 10% of the world's total electrical power) [3]. *One of the messages of this book is that nuclear (fission plus fusion) could provide all the world's current electricity for millennia.*

Nuclear power is reliable. The median capacity factor—the ratio of the actual 809.4 terawatt-hour(e) = 8.1×10^{11} kWh(e), kilowatt-hour(e), of nuclear (fission) electrical energy produced in the year by the 98 US nuclear reactors, divided by the theoretical amount of energy that could have been produced if these reactors had all run at 100% design power for every second in the entire year – was an impressive 91% in 2017–19 [4]. Clearly, nuclear power is a reliable, carbon-free, and mature technology for large-scale electric power production, which could and should be more extensively employed on a worldwide scale, now, to displace power production by burning carbon-based fuel and thereby avoid carbon burning's associated impact of climatic degradation, and thus prevent, or at least lessen, mankind's impending environmental catastrophe [2].

A major worldwide expansion of nuclear power to displace carbon-based power requires (1) adequate availability and more efficient technical utilization of the nuclear fuel resources; (2) an expansion and perhaps some reorganization of the skilled workforce for designing, manufacturing, constructing, and operating new nuclear power plants and the supporting industrial plants on a major scale; and (3) a worldwide political leadership committed to actively engaging climate degradation, which involves dealing with the financial interests of carbon-based fuel, among others. This book first discusses nuclear physics and technology and nuclear fuel resources, issues of nuclear power, with an emphasis on the unique role that can be played by subcritical nuclear fission reactors with fusion neutron sources, known as Fusion–Fission Hybrid (FFH) reactors. Then the FFH is analyzed in detail.

Chapter 3
Scientific basis of nuclear fission energy

The scientific basis for nuclear fission power is the conversion of a small amount of mass to an enormous amount of energy in nuclear fission reactions. For atoms of some isotopes of uranium and heavier transuranic (TRU) elements, the addition of a subatomic particle known as neutron to the nucleus results either in (1) the *fission* of the original nucleus into two or three nuclei plus two to three neutrons, less massive in sum, or (2) the *capture* of the neutron to create a nucleus one atomic mass unit heavier, or (3) the *scatter* of the original neutron to generally produce a more energetic nucleus and a less energetic neutron. (Neutrons may also scatter from an atom without entering the nucleus.) The different probabilities used for neutron capture, scattering, and fission happen to have the units of area and are thus colloquially known as "cross sections" for these neutron–nucleus reactions. In the case of neutron *capture* by a nucleus it is thereby *transmuted* into a nucleus one atomic mass unit (A) heavier but with the same nuclear charge (Z), the process being denoted $[(A, Z) + \mathrm{n} \Rightarrow ((A + 1), Z)]$. In the case of *scatter,* the nuclear constituents, hence the mass, remain unchanged, but the kinetic and internal energies of the nucleus are generally increased while the kinetic energy of the neutron is generally decreased.

In the case of *fission*, the total mass of the resulting *fission product* nuclei plus neutrons is less than the mass of the original nucleus plus the mass of the neutron causing the fission, and this difference in mass is converted into about 200 MeV (million electron-volts) kinetic energy of atomic and nuclear particles per fission, which is converted to thermal energy in the immediate proximity of the fission reaction, recovered, and converted to electricity.

The neutrons released in the fission event may escape (*leak*) from the reactor, be *captured* by or *scattered* from another uranium atom or structural, coolant, or control material atom in the reactor, or cause *fission* in another uranium or plutonium atom, producing a further two to three neutrons, as depicted in Figure 3.1.

An enormous number of such reactions go on essentially simultaneously in a nuclear reactor. If, on average, exactly one of the two to three neutrons released in each fission event produces another fission event, *a self-sustaining neutron chain fission reaction* results and the reactor is said to be "*critical.*" If, on the other hand, less or more than one of the two to three neutrons released in the fission event on average produces another fission event, the reactor is "*subcritical*" or "*supercritical,*" respectively. The total neutron population remains constant in time in a critical reactor, increases in time in a supercritical reactor, and decreases in time in

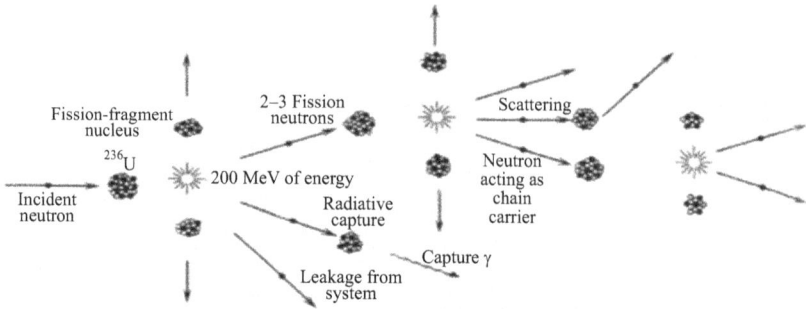

Figure 3.1   Schematic of a fission neutron chain reaction [5]

a subcritical reactor. Nuclear reactors are usually designed to operate *"critical,"* so that the neutron population remains constant in time, but they can readily be designed to operate subcritical by providing a neutron source to maintain a constant neutron population in time.

For a given type of nucleus named X with Z protons and $A-Z$ neutrons, the standard designation is $^{A}X_{Z}$ and the chemical name X is related to Z (e.g., for $Z = 6$, X is carbon "C"; for $Z = 92$, X is uranium "U"). Atoms with the same number of protons Z but a different number of neutrons and hence a different number of neutrons plus protons, A, are referred to as *"isotopes"* of element Z (e.g., $^{238}U_{92}$ and $^{235}U_{92}$ are isotopes of uranium with the same number of protons, $Z = 92$, but with different numbers of neutrons and hence different atomic masses, $A = 235$ and 238).

If, on average, exactly one of the two to three neutrons released in the fission event produces another fission event, *a self-sustaining neutron chain fission reaction* results and the reactor is said to be *"critical."* It is also possible to maintain a sustained neutron chain reaction in a subcritical reactor in which on average less than one of the two to three neutrons released in the fission event causes another fission, by the addition of extra neutrons from an external source some of which that do produce a fission event, so that the fission chain reaction is sustained by a combination of fission neutrons + source neutrons in a *"subcritical"* reactor. At the time of the development of nuclear reactors in WWII, neutron sources of the requisite strength did not exist and it was necessary to operate nuclear reactors "critical." This is no longer the situation.

The magnitudes of the capture and fission cross sections are strongly dependent on the energy of the neutron entering the nucleus relative to the excitation energies of excited nuclear states that may thereby be formed.

The neutron fission cross sections of the strongly fissionable isotopes of uranium $^{233}U_{92}$, $^{235}U_{92}$ and of plutonium, $^{239}Pu_{94}$, are shown in Figure 3.2. For these three principal fissionable isotopes, $^{233}U_{92}$, $^{235}U_{92}$, $^{239}Pu_{94}$, and for the also fissionable $^{241}Pu_{94}$, the fission cross section for *thermal* (energy about 10^{-2} eV) neutrons is about $1000 = 10^{3}$ *barns* (1 barn $= 10^{-24}$ cm^2), decreasing with increasing neutron energy to about 1 barn at 1 MeV$=10^{6}$ eV neutron energy, as shown in Figure 3.2. There is a strong "resonance structure" (large, fine-structure variation of cross-section magnitude

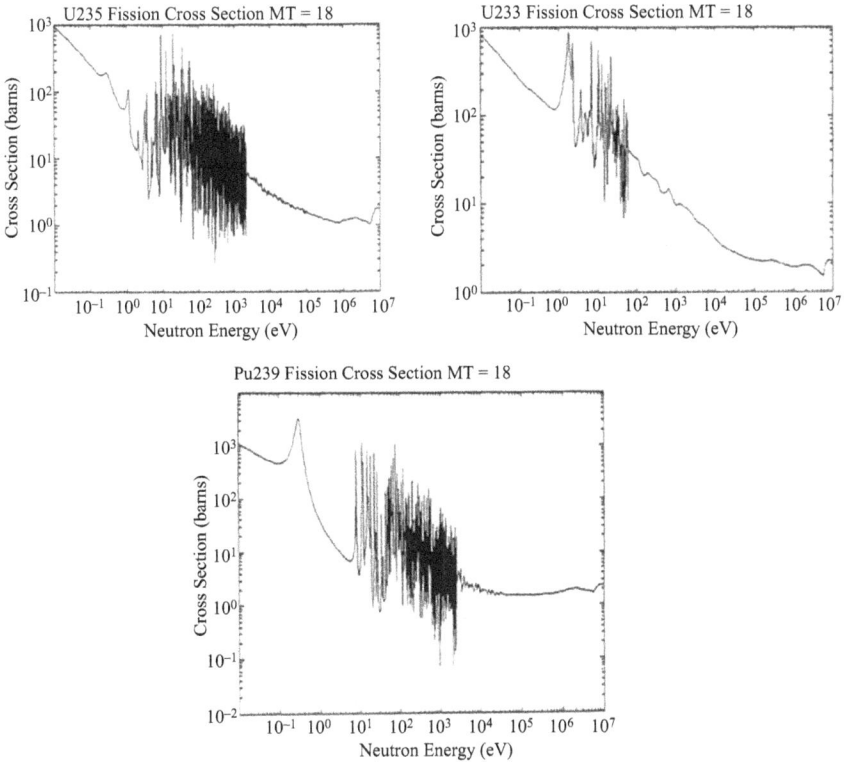

Figure 3.2   Neutron fission cross sections [5]

with neutron energy) in the 1–1,000 eV range, arising from the neutrons with energies corresponding to the many possible excited energy states of the nucleus that can be formed upon the capture of a neutron with the right energy. This energy dependence of the cross sections causes a difference in the physics characteristics of reactors with different neutron energy distributions, as discussed in [5] and elsewhere.

Fission neutrons are released with an energy distribution that broadly peaks in the 1 million eV, or 1 MeV, range, as shown in Figure 3.3.

The first question with regard to the feasibility of maintaining a neutron chain fission reaction is related to how many neutrons are produced in a fission event and how many of these are lost by parasitic capture in non-fissionable nuclei or leakage from the reactor.

The parameter η:

$$\eta \equiv \nu \frac{\sigma_f}{\sigma_a} = \nu \frac{\sigma_f}{\sigma_f + \sigma_\gamma} = \frac{\sum_j \nu_j N_j \sigma_{fj}}{\sum_j N_j (\sigma_{\gamma j} + \sigma_{fj})} \tag{3.1}$$

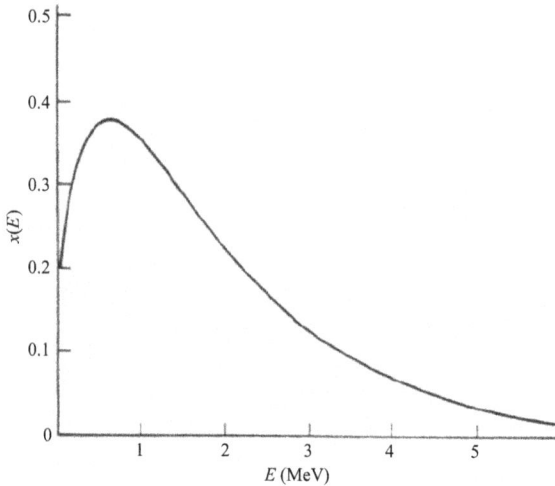

Figure 3.3   Fission spectrum (fission neutron energy distribution) [5]

where v is the average number of neutrons released in a fission event (in the range two to three), σ_f is the fission cross section, and σ_γ is the capture (w/o fission) cross section, and the sum is over all atomic species with number density N in the reactor (or local region of the reactor), provides a measure of the ratio of the fission production rate of neutrons to the fission plus capture loss rate of neutrons in the reactor. Neglecting neutron leakage for the moment, $\eta = 1$ corresponds to a self-sustaining fission chain reaction, so taking neutron leakage into account, $\eta > 1$ is required for a self-sustained fission chain reaction. If η is evaluated just for the fissionable atoms, it is somewhat greater than 2 at very low energy (10^{-3} eV) and has widely varying "resonance" structure with respect to neutron energy at intermediate neutron energies (Figure 3.4). For a fast neutron reactor the $(n, 2n)$ reaction should be included in the numerator of Eq. (3.1).

Both the fission and capture probabilities (cross sections) and the parameter η depend strongly on the energy of the neutron reacting with the nucleus and on the distributions of excited nuclear states for that particular nuclear species. The distribution of neutron energies within a nuclear reactor depends strongly on the materials out of which the reactor core is constructed and on how these materials are arranged within the reactor. In nuclear reactors that contain a large number of low atomic number nuclei (H, D, C), collisions will quickly moderate the neutron energy to the thermal energy of the reactor materials (a fraction of an eV). Such reactors are referred to as "*thermal reactors*" (pressurized water reactors (PWRs), boiling water reactors (BWRs, known jointly as light water reactors, LWRs), heavy water D_2O reactors (CANDUs), and graphite (carbon)-moderated gas-cooled reactors (GCRs), etc. [5]). Nuclear reactors that do not contain many, if any, low atomic mass isotopes will moderate the neutrons only somewhat to a "*fast reactor*" spectrum peaking in the 1,000 eV, or keV, range. Figure 3.5 shows the *neutron flux*

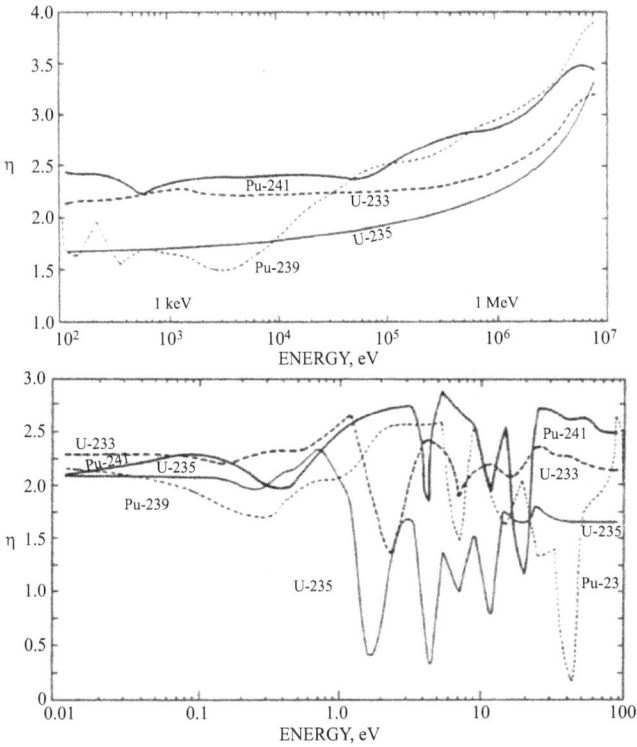

Figure 3.4   The parameter η for the principal fissionable isotopes [5]

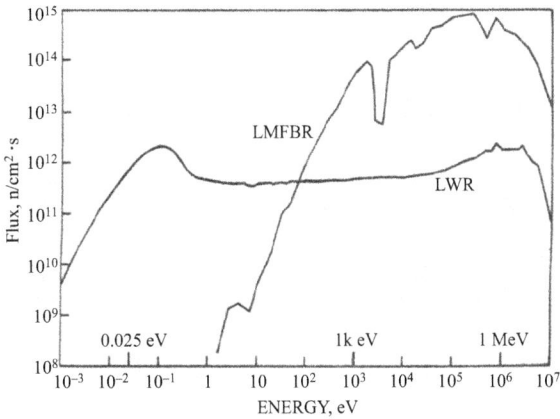

Figure 3.5   Neutron energy spectra in thermal (LWR) and fast (LMFBR) reactors [5]

(the product of the average neutron speed and the number of neutrons per unit volume—the neutron density), calculated for a typical light water *thermal* reactor (LWR) and for a representative liquid metal-cooled *fast* breeder reactor (LMFBR). Thermal reactors and fast reactors will have different physics properties because the average energies of the neutron having reactions, hence the outcomes of the reactions, is very different in the two [5].

For many of the uranium and heavier $A > 92$ TRU isotopes the ratio of the fission to capture cross sections is much larger for a fast reactor neutron energy distribution than for a thermal reactor neutron energy distribution. This energy dependence of cross sections, which is very important for the operation of nuclear reactors, is described in detail in [5], and elsewhere.

Many of the "fission products" (intermediate mass atoms formed in the fission event) are unstable against radioactive decay, emitting a neutron and/or charged particles (proton, alpha (helium nucleus)) from the nucleus (e.g., $(Z, A) \Rightarrow (Z - 1, A - 1)) + proton(1, 1)$ reduces both the charge and mass of the atomic nucleus by one unit).

The radioactive decay rate of an unstable nucleus is characterized by a decay constant $\lambda(Z \rightarrow Z', A \rightarrow A')$ for each possible radioactive decay $(Z \rightarrow Z', A \rightarrow A')$, which constitutes a loss rate $-\lambda_{Z \rightarrow Z-1, A \rightarrow A-1} N_{Z,A}$ for the nuclei species with (Z, A) and a gain rate for the nuclei species with (Z', A'). The "decay product" nucleus (e.g., for proton decay $(Z-1, A-1)$) will in general have different capture, fission, and scattering cross sections than the original (Z, A) nucleus.

Thus, there are many complex *neutron transmutation/radioactive decay chains* in nuclear reactors. Two important such chains, the first of which produces fissionable uranium isotopes ($^{233}U_{92}$ and $^{235}U_{92}$) from essentially non-fissionable thorium atoms ($^{232}Th_{90}$) and the second of which produces fissionable plutonium ($^{239}Pu_{94}$ and $^{241}Pu_{94}$) isotopes from almost non-fissionable uranium ($^{238}U_{92}$), are shown in Figure 3.6. The second of these neutron transmutation/radioactive decay chains is the key that unlocks 99% of the potential nuclear energy resources contained in the world's uranium ore, and the first decay chain is the key that unlocks all the potential nuclear energy content of the world's thorium ore.

Horizontal arrows in Figure 3.6 indicate changes due to neutron capture; vertical arrows indicate changes due to radioactive decay (times shown are radioactive decay half-lives, $t_{1/2} = \ln 2/\lambda$); dashed diagonal arrows represent fission events. Numbers without units are values of the cross sections in units of barns $= 10^{-24}$ cm^2.

The key to both fissile isotope production ("breeding") chains in Figure 3.6 is a plentiful source of neutrons adequate both to maintain the fission chain reaction that produces the nuclear power and to drive the transmutation/radioactive decay chains that convert non-fissionable $^{232}Th_{90}$ or $^{238}U_{92}$ into fissionable uranium or plutonium, respectively, isotopes.

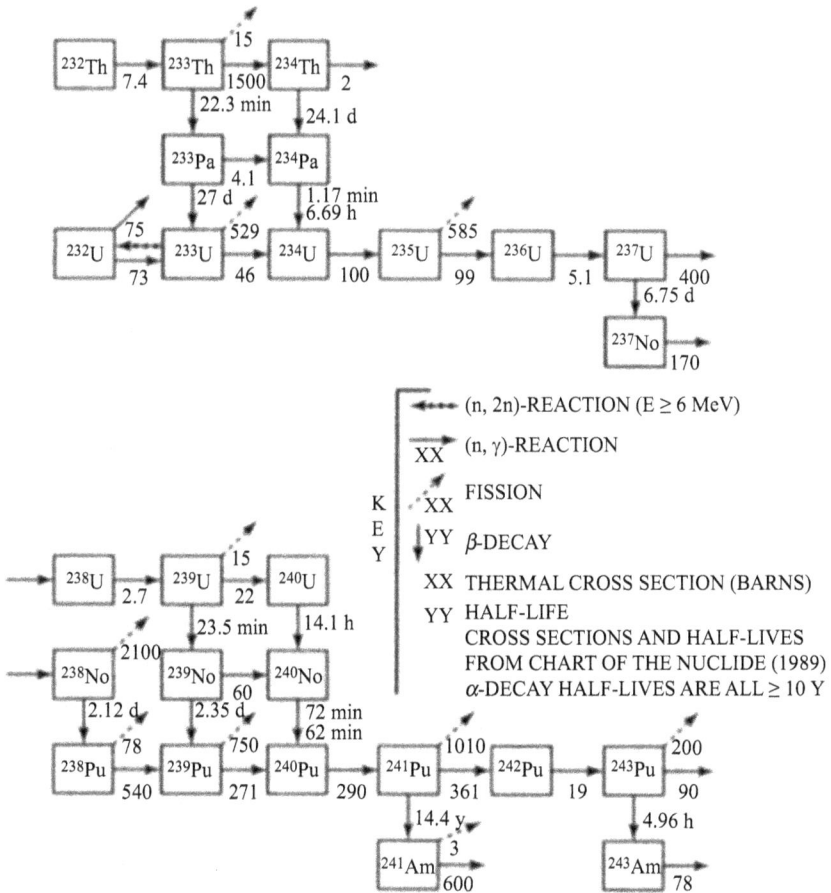

Figure 3.6   Transmutation/radioactive decay chains for $^{232}Th_{90}$ and $^{238}U_{92}$ [5]

Chapter 4

Uranium nuclear fission power fuel cycle

The uranium nuclear fuel cycle is the sequence of industrial processes involved in the production of electricity from uranium and disposing of the waste. While different reactor types and reactors follow somewhat different processes involving somewhat different masses of different materials, the generic process and material balance provided by the World Nuclear Association [6] for a representative 1,000 MWe PWR using 4.5% enriched fuel to achieve 45 gigawatt-day/ton (GWd/t) burnup depicted in Figure 4.1 provides a representative case. However, we note that not all steps (e.g., reprocessing, disposal) are included in the present fuel cycles.

Mining of uranium ore from open pit or underground mines or by circulating oxygenated groundwater through a porous ore body to dissolve the uranium oxide and bring it to the surface is the first step. From 20,000 to 400,000 tons of uranium ore, depending on the uranium concentration, is needed annually for a single 1,000 MWe reactor.

Milling crushes and grinds the mined ore to a fine slurry, which is leached in sulfuric acid to separate the uranium from waste rock; the uranium ore is then recovered and precipitated as 249 tons of uranium oxide (U_3O_8) concentrate ("yellowcake") containing 211 tons of uranium.

Conversion of the uranium oxide results in 312 tons of uranium hexafluoride (UF_6) gas, which is sent to the *enrichment* plant (previously gaseous diffusion plants, but now a bank of cyclotrons), where the (UF_6) gas is separated into two streams, one containing 24.3 tons of enriched uranium at 4.5% U235 ("low enrichment uranium") and the other depleted from 0.72% to 0.22% in U235 (the "tails").

Fuel fabrication is generally into the form of pressed UO_2 ceramic pellets which are baked at temperatures $> 1,400°C$, then encased in metal tubes to form fuel rods that are assembled into "fuel assemblies." About 27 tons of new enriched uranium fuel is required annually by a 1,000 MWe nuclear reactor.

Fuel burnup and power generation. Several hundred assemblies containing about 75 tons of low enriched (4.5% in this example) uranium make up the core of a 1,000 MWe PWR, and the U235 partially undergoes neutron fission to produce about 200 MeV of ultimately thermal energy per fission. In addition, some of the U238 is turned into fissionable Pu239 and Pu241 in the second neutron transmutation/radioactive decay chain shown in Figure 3.6. The Pu239 and Pu241 are also partially fissioned to provide about one-third of a PWR's or a BWR's energy production (about one-half in CANDU reactors).

The Nuclear Fuel Cycle

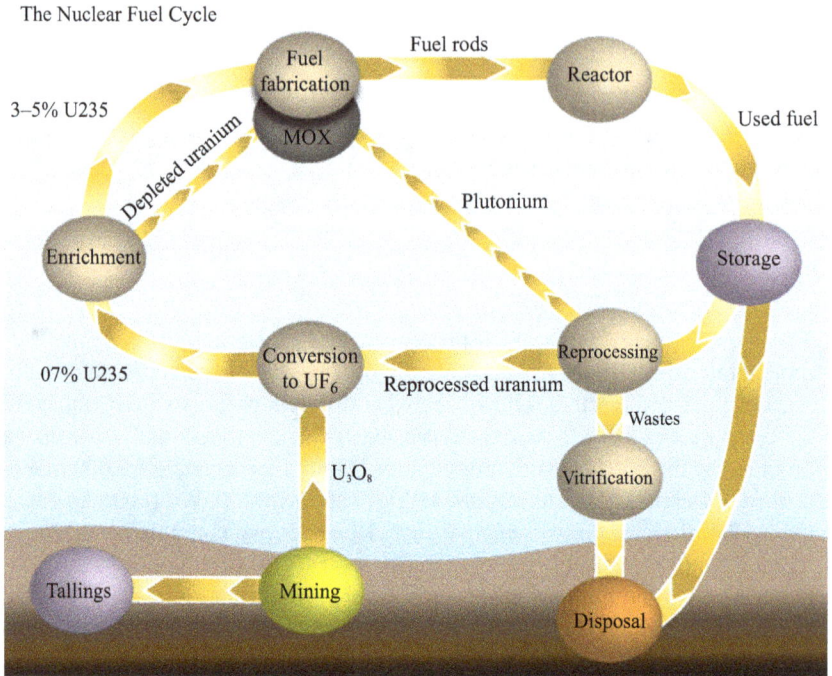

Figure 4.1   The nuclear fuel cycle for uranium [6]

About one-third of the fuel in a reactor is removed and replaced with fresh fuel every year to 18 months, in order to maintain efficient reactor performance.

Typically, about 44 million kWh of electricity is produced from 1 ton of natural uranium. (The production of this amount of electricity from fossil fuels would involve burning over 20,000 tons of black coal or 8.5 million m^3 of gas.)

Used or *spent fuel* removed from a reactor will typically be about 95% U238, 1% U235, 0.6% fissile Pu, and 3% neutron-absorbing fission products and minor actinides.

Reprocessing and fissionable fuel recycling

The used fuel removed from reactors still contains about 96% of its original uranium (with a slightly higher U235 fraction than in natural uranium), about 1% fissionable Pu, and about 3% parasitic neutron-absorbing fission products and higher A ($Z > 92$) transuranics. Removing the parasitic neutron absorbers and recycling the uranium and plutonium as fuel for subsequent reactors would utilize a larger fraction of the potential energy of uranium. To a limited extent, this is done today in *mixed oxide (MOX) fuels*.

The different reprocessing technologies which separate metals from their mineral concentrate—pyrometallurgy (heat), electrometallurgy (electricity), and hydrometallurgy (water)—are discussed in [6].

In most present nuclear fuel cycle options, the *spent* nuclear fuel is not reprocessed but rather placed in temporary storage, usually at the reactor site, awaiting either future reprocessing and/or disposal in high-level waste repositories (HLWRs). Such fuel cycles are known as "once-through" fuel cycles. *Without reprocessing, such "once-through" fuel cycles would ultimately rebury 96% of the potential fission energy content of the uranium nuclear fission fuel resources plus the 1% fissionable plutonium resource created by neutron capture in the used fuel, or rebury 97% of the original mined uranium fuel potential energy resource.* However, the intention is to later reprocess this "spent fuel" that is currently being stored, for the purpose of recovering and using for fuel the plutonium and unused uranium, thereby gaining 25–30% more energy from it, as well as reducing the volume by a factor of 5 and the radioactivity level substantially of spent fuel sent to the HLWRs, after 100 years of radioactive decay.

New "generation-4" fast reactors have more favorable characteristics for fissioning the used fuel from the present thermal spectrum reactors than do the present thermal spectrum reactors themselves, which would not only enable the reprocessing not only of used fuel from present reactors to become fuel for new reactors, but also enable the large stockpiles of depleted uranium from enrichment plants (1.5 million tons by 2015) to become a fuel source.

Another major advance would be the possibility of recovering all long-lived actinides (actinium with $Z = 89$ to americium with $Z = 103$, including plutonium) and recycling them to extract fission energy while transmuting the long-lived fissionable waste (with 100,000+ year half-lives) into fission products with much shorter (approximately 100-year) half-lives. *Between 2010 and 2030, there will be 400,000 tons of used nuclear fuel generated, and this could be converted to an enormous nuclear fuel resource, rather than buried.*

Reprocessing nuclear fuel cycle options, which would utilize some, if not all, of this potentially wasted 97% of the uranium energy resource in used nuclear fuel and the depleted uranium, are known as "breeding" fuel cycles because they convert non-fissionable U238 to fissionable Pu239 and Pu241 and heavier transuranics, as shown in the second chain in Figure 3.6, and fission them to produce about 200 MeV of energy per fission. The development of such fuel cycles has the potential to increase the recoverable energy of the uranium nuclear fuel resources by two orders of magnitude relative to the present "once-through" fuel cycle. *The key to the development of such "breeding" fuel cycles for uranium is the availability of sufficient neutrons to drive the second transmutation/radioactive decay chain in Figure 3.6, as well as maintain the neutron fission chain reaction. A thesis of this book is that fusion could provide these neutrons.*

Chapter 5
Fission energy fuel resources

The isotope $^{235}U_{92}$ has a large fission cross section for thermalized (very low energy) neutrons (Figure 3.2), but $^{235}U_{92}$ constitutes only 0.72 atom% (0.71 wt%) of natural uranium, the remaining 99.3% of which is $^{238}U_{92}$, which has a negligibly small "thermal" fission cross section and a small but non-negligible fast fission cross section. However, *neutron capture in the more plentiful $^{238}U_{92}$ initiates a variety of transmutation/radioactive decay chains that convert the almost non-fissionable $^{238}U_{92}$ into various transuranic isotopes [5], some of which have quite large fission cross sections, notably $^{239}Pu_{94}$ and $^{241}Pu_{94}$, the creation process of which is illustrated in the second decay/transmutation chain as shown in Figure 3.6.*

5.1   Uranium resources

The world's estimated natural uranium (0.72% $^{235}U_{92}$ and 99.28% $^{238}U_{92}$) and thorium (100% $^{90}Th_{232}$) ore resources are given in Tables 5.1 and 5.2, respectively (in units of MT = metric ton). The cited numbers represent identified resources (as of January 1, 2017) recoverable at a cost of up to US $130/kg of U or Th. There are additional resources identified at higher extraction costs and/or available in non-traditional sources (e.g., ocean water). Further resources are also likely to be discovered by future exploration.

Energy is extracted from natural uranium via (1) direct neutron fission of the 0.72% $^{235}U_{92}$ in natural uranium and by (2) neutron capture in $^{238}U_{92}$ followed by transmutation and radioactive decay chains leading from $^{238}U_{92}$ to $^{239}Pu_{94}$ or $^{241}Pu_{94}$, followed by neutron fission of the resulting $^{239}Pu_{94}$ or $^{241}Pu_{94}$ atoms, and by further neutron capture/radioactive decay transmutation reactions leading to higher mass fissionable transuranic (TRU) isotopes. Figure 3.6 illustrates the neutron transmutation/radioactive decay chains involved.

The amount of nuclear energy actually extracted depends upon the fuel cycle employed, whether it is "economically optimized" to minimize cost or "resource optimized" to maximize energy output, or somewhere in between. Present nuclear reactors use an inefficient "once-through" fuel cycle, in which the natural uranium is first "enriched" (previously in massive gaseous diffusion plants, but now using banks of centrifuges) in the fissionable $^{235}U_{92}$ isotope from the 0.72% in natural uranium ore to 4–6% in LWR reactor fuel, in order to facilitate maintaining a neutron chain fission reaction for 1–2 years, after which point the partially utilized

*Table 5.1   World's estimated natural
uranium resources in MT
(IAEA-NEA 2017 [7])*

Australia	1,818,300
Kazakhstan	842,200
Canada	514,400
Russia	485,600
Namibia	442,100
South Africa	322,400
China	290,400
Niger	280,000
Brazil	276,800
Uzbekistan	139,500
Ukraine	114,100
Mongolia	113,500
Botswana	73,500
Tanzania	58,200
United States	47,200
Jordan	43,500
Other	280,600
Total	6,142,600

fuel is discharged from the reactor. In the widely used "once-through" fuel cycle the discharged fuel is placed in temporary spent fuel storage to await ultimate burial in high-level waste repositories (HLWRs), before or after reprocessing to recover the remaining fissionable material.

According to [7], annual world reactor-related uranium usage in 2016 (in the "once-through" fuel cycle) was 62,825 tons U. Thus, at the current nuclear electricity production level and with use of the current "once-through" fuel cycle (which utilizes only 1–3% of the uranium energy content), identified uranium resources could supply the world's present (as of 2016) *nuclear electricity* production rate for about 100 years. Because of new reactors being built (primarily in Asia), the world's nuclear electricity production rate and the corresponding required uranium resources are expected to moderately increase, but in all likelihood so will the identified resources. This is based on the present (climatically disastrous) "business as usual" scenario in which burning increased levels of carbon-based fuels continues to provide most of the world's new electricity production.

We can make a rough calculation of the present energy extraction rate per ton from natural uranium (with the widely used "once-through" fuel cycle). Natural uranium has 0.72% $^{235}U_{92}$, and current PWRs use 4–5% enriched fresh fuel, say 4.55%, which translates into the necessity to concentrate the $^{235}U_{92}$ in the natural uranium ore about 6.5 times. Taking into account that some of the $^{235}U_{92}$ is lost in the "tails," it is estimated that about 8 MT of natural U is required to produce 1 MT of uranium enriched to 4.55%. The achievable discharge burnup of fuel with such enrichment should be about 60,000 MWd (thermal)/MTU, depending on reactor design, reloading strategy, number of batches of fuel, etc. At *present thermal-to-electrical conversion efficiencies* this is about 480,000,000 kWh(e) per MT of enriched uranium. For 8 MT of natural uranium this *is* 60,000,000 *kWh(e) per MT of natural uranium.* (Similar estimates of mass–energy conversion efficiencies in the "once-through" fuel cycle are found in the literature.) The world's estimated uranium resources are given in Table 5.1. Using the above conversion ratio of 60 million kWh(e) per MT of natural uranium, which corresponds to the present "once-through" fuel cycle, the world's estimated uranium resources would produce 3.68×10^{14} kWh(e) of electricity if utilized in the "once-through fuel cycle."

The entire annual world production of electricity (from all fuel sources) in 2017 was 2.14×10^{13} kWh(e). What it will be in the future depends on a number of unknown factors. If we simply assume that the world's annual total electricity production rate remains at the 2017 level, we can calculate from this line of argument that *there is enough uranium to provide the entire present level of world electricity production for a couple of decades, using the present "once-thru" fuel cycle,* which fissions most of the $<1\%$ $^{235}U_{92}$ in the original natural uranium and utilizes some small fraction of the $^{238}U_{92}$ that is transmuted into fissionable plutonium isotopes and then fissioned.

The nuclear energy available from uranium using the "once-through" cycle is an enormous amount of energy, of course, but only a small fraction ($\approx 2\%$) of the potentially available energy from the world's uranium ore resources, when we take into account that much, if not all, of the non-fissionable $^{238}U_{92}$ (which constitutes 99.3% of the natural uranium) could be converted in a "breeding" fuel cycle into fissionable $^{239}Pu_{94}$ or $^{241}Pu_{94}$ or higher mass transuranics by neutron capture in $^{238}U_{92}$ followed by the second transmutation/radioactive decay sequence shown in Figure 3.6. Fuel cycles which focus on the production of $^{239}Pu_{94}$ and $^{241}Pu_{94}$ by neutron capture in $^{238}U_{92}$ followed by the second of the transmutation/radioactive decay sequences as shown in Figure 3.1, followed by the neutron fission of this $^{239}Pu_{94}$ or $^{241}Pu_{94}$, in the same or other reactors, are known as "breeding" fuel cycles. There is, of course, some small amount of breeding of fissionable $^{239}Pu_{94}$ and $^{241}Pu_{94}$ by neutron capture in $^{238}U_{92}$ followed by fission of the Pu in the "once-through" cycle, but the cycle certainly is not optimized for that purpose. *The possibility of a "breeding" fuel cycle is very dependent on the number of neutrons, in addition to those required to maintain the fission chain reaction, that are available to breed new fissionable isotopes.*

In principle, a much larger fraction of uranium atoms could be fissioned, directly in the case of $^{235}U_{92}$, and after conversion of $^{238}U_{92}$ to $^{239}Pu_{94}$, $^{241}Pu_{94}$, or higher fissionable transuranics in the case of $^{238}U_{92}$, to release about 200 MeV per atom of U fissioned, if the fuel can be exposed to an adequate neutron flux for long enough and the reactor can be maintained critical or operated subcritical and supported by an adequate neutron source. Complete fissioning of every uranium atom, either directly or after transmutation(s) into a fissionable atom, would produce about 22.5×10^9 kWh(t) of thermal energy per ton of natural uranium. With advanced thermal-to-electrical conversion efficiencies of about 45%, 1 ton of natural uranium could theoretically produce about 10^{10} kWh(e) of electrical energy. In reality, processing losses, fuel limitations, and other technical constraints would likely reduce this ***theoretical upper limit to 4–5 $\times$ 10^9 kWh(e) per ton of natural uranium.***

Multiplying this last number by the 6 million tons total natural uranium estimate shown in Table 5.1 would indicate that the world's estimated uranium resources could potentially provide the world's present electricity production for ~1000 years if utilized in an efficient "breeding" fuel cycle. This estimate of about 1000 years of the world's current electricity production for a perfect breeding fuel cycle which fissions all the uranium atoms, while arguably overly optimistic, is consistent with our previous estimate of several decades (tens of years) of the world's current electricity production for the once-through fuel cycle, which fissions only the about 1% of the uranium atoms that are $^{235}U_{92,}$ plus a small fraction of the $^{238}U_{92}$ atoms that are neutron-transmuted into fissionable $^{239}Pu_{94}$ or $^{241}Pu_{94}$. Actually being able to fission all the uranium is highly improbable for a number of practical reasons: the fuel clad fails after a certain amount of neutron irradiation (about 200 displacements per atom is the best current estimate) and the fuel must be reprocessed to remove the neutron-absorbing fission products and to recover and refabricate the still fissionable uranium and plutonium into fuel elements, processes with some inevitable loss involved, which are necessary to maintain an efficient fission chain reaction.

Furthermore, the ability to maintain a critical reactor with additional neutrons being available for this conversion (breeding) of $^{238}U_{92}$ into $^{239}Pu_{94}$ and $^{241}Pu_{94}$ and higher transuranics is an issue. (Note that the challenge of maintaining an adequate neutron flux for both breeding fissionable isotopes and maintaining the neutron chain fission reaction would be greatly facilitated by (and probably requires) an external neutron source (e.g., D–T fusion [9–14] or accelerator-spallation [16] are currently the only technically credible possibilities).)

The difference between the theoretical upper limit estimate of a thousand years of the world's current nuclear electricity production for a future "breeding" fuel cycle, on one hand, and the upper limit of decades for the present "once-through" fuel cycle calculated above certainly defines a near term "grand challenge" for nuclear power, but a credible challenge. Rather than complicating this discussion further with a comparison of various fuel cycle scenarios that have been examined [16], we will simply leave it ***that the world's known uranium resources could supply the present total world electricity production for a few decades (based on the present "once-through" fuel cycle) to a few thousand years (based on an advanced "breeding" fuel cycle),*** *depending on the efficiency of the fuel cycle employed.* ***Utilizing more than a few percent of the world's potential uranium energy resources would seem to be greatly facilitated by, if not dependent on, the development of an adequate neutron source to "drive" subcritical nuclear fission transmutation reactors.*** *A thesis of this book is that plentiful fusion neutrons could be employed to provide fissionable isotopes by neutron transmutation of* $^{238}U_{92}$.

Two types of technically credible neutron sources have been proposed: D–T fusion and accelerator-spallation. The fusion neutron source would be broadly distributed over the rather large surface area of the fusion plasma chamber, which suggests that the nuclear reactor should be configured to be located adjacent to the surface of the fusion plasma chamber; i.e., the nuclear reactor core should be "wrapped around" the fusion plasma neutron source, which is a credible configuration. On the other hand, the accelerator-spallation neutron source is a very localized "point" source, which suggests the likelihood of a severe radiation damage problem in the vicinity of such a "point source" in a nuclear reactor.

5.2   Thorium resources

The other highly fissionable isotope of uranium is $^{233}U_{92}$, which is produced by neutron capture in thorium, $^{232}Th_{90}$, followed by two successive radioactive decays (see Figure 3.6). The world's estimated thorium reserves are given in Table 5.2. Energy would be extracted from thorium in a similar "breeding" fuel cycle by neutron capture in $^{232}Th_{90}$ followed by radioactive decay and neutron transmutation chains leading from $^{232}Th_{90}$ to $^{233}U_{92}$ or to $^{235}U_{92}$, both of which are neutron fissionable.

This thorium would have to first be irradiated with neutrons to produce fissionable $^{233}U_{92}$, presumably at the same 4–5% as the enrichment of $^{235}U_{92}$ in uranium fuel, if a critical reactor is to be used to fission this fuel. It is not clear at this time whether this "breeding" would be done in situ in the same reactor in which the fuel would subsequently be fissioned or in special fuel cycle preprocessing "breeder" reactors, or in both.

Table 5.2   World's estimated thorium resources in MT (IAEA-NEA 2014 [8])

India	846,000
Brazil	632,000
Australia	595,000
United States	595,000
Egypt	380,000
Turkey	374,000
Venezuela	300,000
Canada	172,000
Russia	155,000
South Africa	148,000
China	100,000
Norway	87,000
Greenland	86,000
Finland	60,000
Sweden	50,000
Kazakhstan	50,000
Other	172,000
Total	6,355,000

More or less the same "resource utilization" logic applies to thorium as discussed above for natural uranium. In principle, almost every thorium atom could be fissioned after conversion to $^{233}U_{92}$, $^{235}U_{92}$, or higher fissionable transuranics to release about 200 MeV per atom of $^{232}Th_{90}$, if the fuel can be exposed to an adequate neutron flux long enough and if the reactor can be maintained critical or supported by an adequate neutron source. Complete fissioning of every thorium atom, after transmutation(s), would produce about 22.5×10^9 kWh(t) of thermal energy per ton of thorium. With advanced thermal-to-electrical conversion efficiencies of ~45%, 1 ton of $^{232}Th_{90}$ ore could theoretically produce about 10^{10} kWh(e). In reality, as with uranium, processing losses, fuel limitations, and other technical limits would likely reduce this theoretical upper limit to 4–5×10^9 kWh(e) per ton of $^{232}Th_{90}$. Multiplying this number by the ~6 million tons total thorium ore estimate shown in Table 5.2 would indicate *that the world's known thorium resources could supply the present world electricity production level for a few thousand years, depending on the efficiency of the "breeding" fuel cycle employed. Such a breeding fuel cycle would certainly be facilitated by an external neutron source, and the advantages of a fusion neutron source discussed for uranium also pertain to thorium.*

Chapter 6

Technological basis of nuclear fission power

The original man-made (*there is evidence of a prehistoric nuclear reaction in an African cave*) nuclear reactor constructed in Chicago during the WWII Manhattan Project, under the leadership of Enrico Fermi, was able to maintain a constant neutron flux level without an external neutron source ("go critical") using natural uranium of the composition of the mined uranium ore by making use of a number of design strategies such as lumping the fissionable isotopes rather than a uniform dispersion. However, it was determined at the time that such reactors were impractical and that there would be great advantage in operational capability and flexibility if the uranium was "enriched" in the highly fissionable isotope $^{235}U_{92}$ and/or if some fissionable plutonium could be bred from $^{238}U_{92}$ (hence the rationale for the path taken in the Manhattan Project).

The process used for this enrichment of the 0.7% fissionable $^{235}U_{92}$ content of natural uranium to about 4–5% must be based on the small mass differences (3 out of 238, or about 1.3%) between the uranium isotopes, which until recently required large gaseous diffusion plants, but now is done with banks of centrifuges, in order to produce enriched uranium fuel for practical nuclear (fission) power reactors.

Nuclear fission power reactors consist of (1) a *core*, an assemblage of fissionable material contained within metal-clad fuel elements, which is capable of maintaining a neutron chain fission reaction for prolonged periods (typically 1–2 years), limited by neutron radiation damage to the fuel materials, depletion of the fissionable material, and accumulation of neutron-absorbing fission product atoms; (2) the fuel elements and the core *structural support system*; (3) a *primary coolant system* which removes fission heat produced in the fuel elements as thermal energy to (4) a *heat exchanger* where the energy is transferred to (5) a *secondary coolant system* to produce steam that (6) drives a *turbine generator to produce electricity*.

The fuel elements for the pressurized water reactors (PWRs) and boiling water reactors (BWRs) typically consist of centimeter-scale radial dimension pins containing uranium, uranium oxide, uranium carbide, or uranium-xxx granules and clad with a metal (frequently zirconium alloy) to provide the primary containment for the fuel and for the radioactive fission products that are formed therein. In gas-cooled thermal reactors the fuel may be in the form of small spherical TRISO pellets contained within multiple layers of silicon carbide. A fuel assembly for a PWR is shown in Figure 6.1.

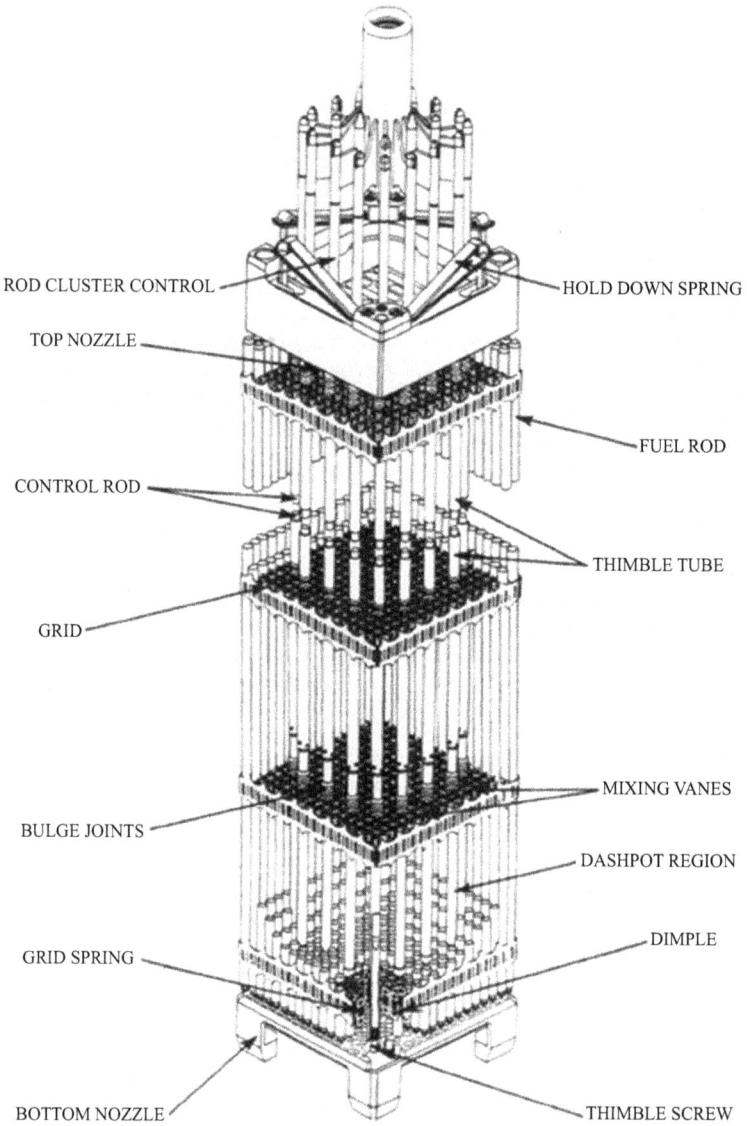

Figure 6.1   Fuel assembly for a Westinghouse pressurized water reactor (PWR) showing fuel rods, control rods, etc. [5]

Most nuclear fission reactors that have been built include low atomic mass "moderators" (to slow the fast fission neutrons down to thermal speeds to take advantage of the much larger fission cross section for "thermalized," than for fast, neutrons). In water-cooled reactors, the coolant also serves as the moderator. In the most common PWR the H_2O is pressurized to about 2,000 psi to avoid boiling. In the BWR, the coolant is at lower pressure, allowing some boiling.

Both PWRs and BWRs today typically use fuel that is 4–6% enriched in $^{235}U_{92}$.

Gas-cooled reactors (HTGRs) operate at higher temperatures than water reactors, use CO_2 or He as coolant, and employ a graphite moderator in order to slow down the neutrons to access the higher thermal fission cross section. These reactors use slightly (4–6%) enriched uranium.

The Canadian CANDU reactor uses D_2O (lower neutron capture cross section than H_2O) in order to be able to go critical with natural (instead of enriched) uranium.

The mean distance that a fission neutron travels before slowing down and being captured or causing a fission is $M \sim 6.6$ cm in a PWR, $M \sim 7.3$ cm in a BWR, and $M \sim 21$ cm in a HTGR. The spatial dimensions (i.e., the core height and core radius) must be large compared to M; otherwise excessive neutron leakage would prevent criticality from being achieved. A typical critical cylindrical PWR core might have dimensions: major radius $R \sim 1.1$ m and height $H \sim 3.7$ m. The dimensions of a similar critical BWR would be expected to be only slightly larger, but the dimensions of a critical HTGR would be about two to three times larger.

The configuration of a typical PWR is shown in Figure 6.2. The core containing the fissioning fuel is located in a pressure vessel in the bottom part of the figure. This core is about 3.5 m in diameter and 3.5–4.0 m high, made up of about 200 of the fuel assemblies shown in Figure 6.1. Each assembly typically consists of a 17×17 array of metal-clad fuel pins about 1 cm in diameter, cooled by H_2O at about 2,000 psi. There are *multiple levels of containment of the radioactive fuel and fission products*, which (1) are contained within *fuel pellets*, which in turn are (2) contained within *metal-clad fuel pins*, which are assembled into the "reactor core," which is contained in a (3) *pressure vessel*, which in turn is contained within a surrounding (4) *containment building* designed to withstand various internal over-pressurization events and external impacts.

The "breeding" or "burning" capabilities of fast and thermal reactors are compared in Figure 6.3. Since a fission reaction produces energy plus two to three neutrons, while a capture reaction only destroys a neutron, the implication of Figure 6.3 is that for essentially all the fissionable isotopes produced in the neutron transmutation/decay chains in Figure 3.6, the probability of fission and the associated neutron production rates are significantly greater in a fast reactor neutron spectrum than in a thermal reactor neutron spectrum. In other words, fast reactors have better "breeding" and "burning" characteristics than do thermal reactors.

Two other water-cooled, thermal reactor types are also widely used. The BWR pioneered by General Electric is similar to the PWR, except for operating at lower coolant pressure, resulting in some core coolant boiling. The Canadian CANDU reactor utilizes pressurized heavy water (D_2O) coolant in a pressure tube configuration and uses natural, rather than enriched, uranium fuel.

A fast reactor concept, the integral fast reactor (IFR) [18], is shown in Figure 6.4. Again, the fissionable fuel is contained in fuel pins located within fuel assemblies in the reactor vessel at the bottom of the figure. This reactor uses a metal fuel in fuel assemblies immersed in a pool of sodium, which is the reactor

CONTROL ROD DRIVE
MECHANISM

THERMAL SLEEVE

CONTROL ROD
DRIVE SHAFT

LIFTING LUG

CLOSURE HEAD
ASSEMBLY

UPPER
SUPPORT PLATE

INTERNALS
SUPPORT LEDGE

HOLD-DOWN SHARING

CORE BARREL

INLET NOZZLE

OUTLET NOZZLE

FUEL ASSEMBLIES

BAFFLE

UPPER CORE PLATE

FORMER

REACTOR VESSEL

LOWER CORE PLATE

LOWER
INSTRUMENTATION
GUIDE TUBE

IRRADIATION
SPECIMEN GUID

BOTTOM SUPPORT
FORGING

NEUTRON SHIELD PAD

RADIAL SUPPORT

TIE PLATES

CORE SUPPORT
COLUMNS

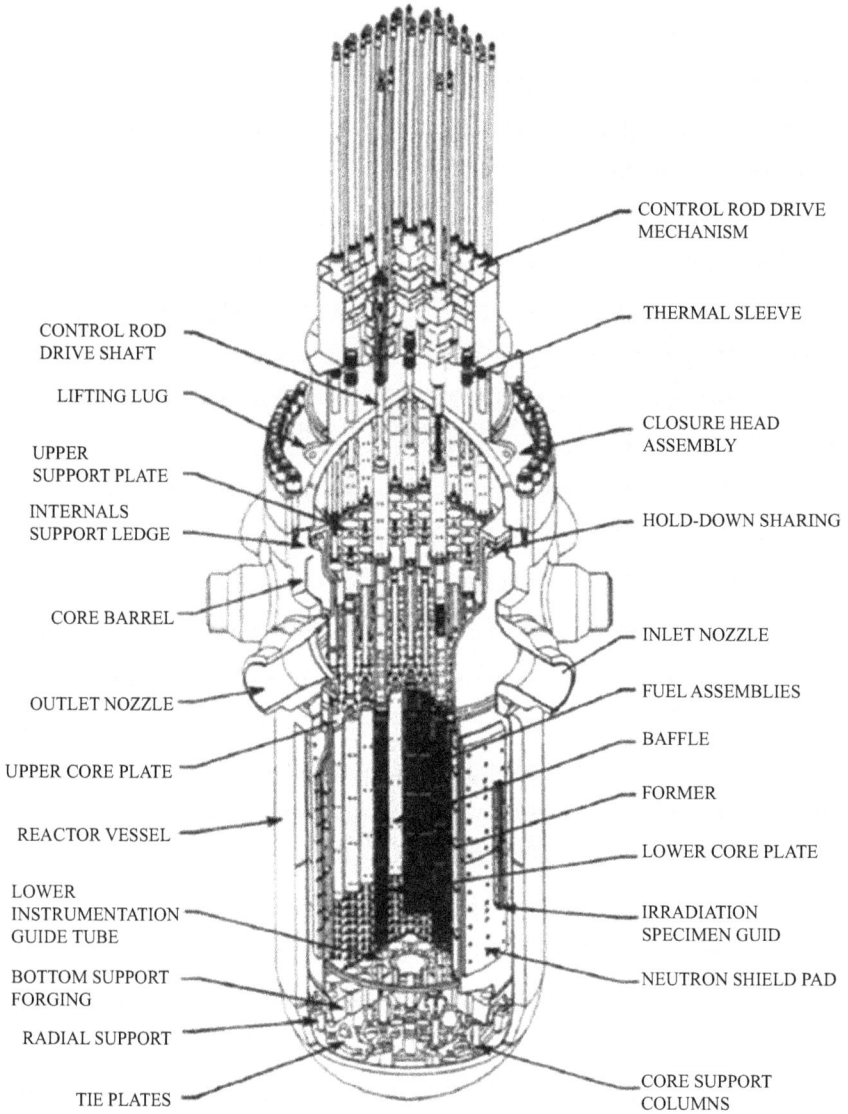

Figure 6.2   A Westinghouse pressurized water reactor (PWR) [5]

coolant. Unlike the PWR, the Na-cooled reactors are not highly pressurized, and the reactor vessel is not a pressure vessel.

An advanced version of this IFR concept is being designed for the Versatile Test Reactor to be built in the United States.

A more extensive description of fission reactor cores can be found in Chapter 7 of [5] and in [18].

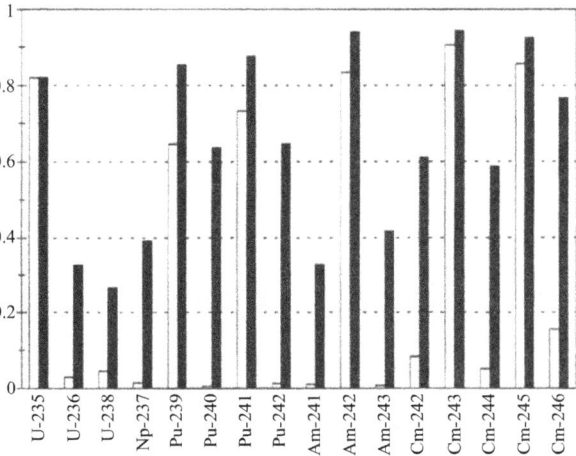

Figure 6.3   Fission likelihood = ratio of the fission to (fission + capture) cross
sections (shaded: fast reactor neutron spectrum; unshaded: PWR
thermal neutron spectrum) [5]

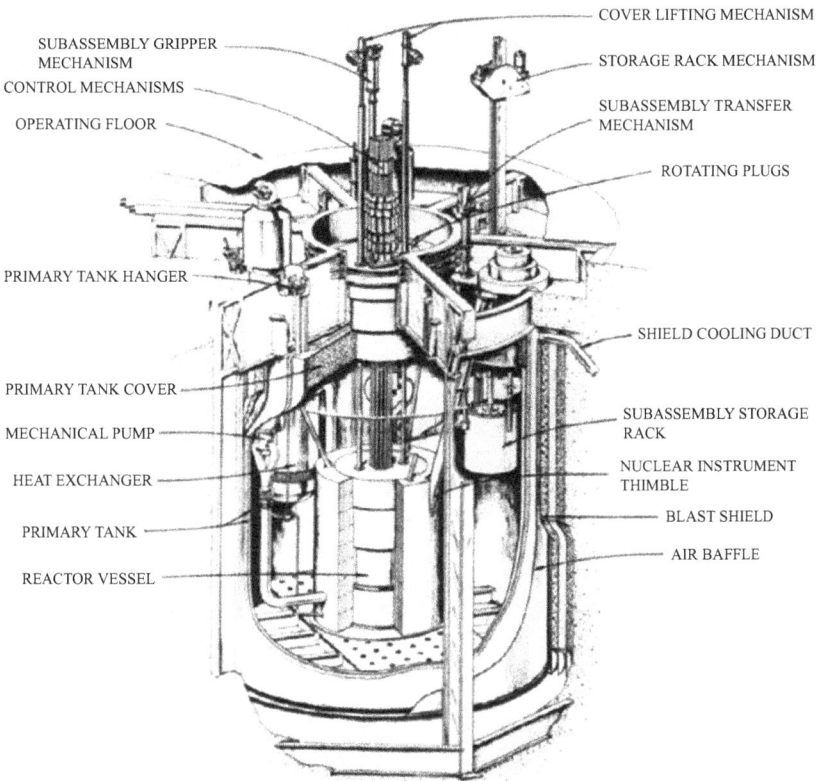

Figure 6.4   Integral fast reactor (IFR) design [17]

A relatively recent summary of power reactors worldwide is given in Table 6.1.

Table 6.1   Power reactors by type, worldwide 2019 [3]

Reactor type	# Units	MWe
Pressurized water (PWR)	370	361,926
Boiling water (BWR)	81	81,766
Gas cooled (GCR)	15	7,885
Heavy water (PHWR)	57	29,836
Graphite moderated (LGR)	15	10,219
Liquid-metal (LMFBR, LMR)	5	1,939
		493,571

Chapter 7

Conversion of nuclear fission energy to electrical energy

The fission of a $^{235}U_{92}$ (or plutonium or other transuranic atom) nucleus with atomic mass $A_{\text{uranium}} = 235$ within a fuel element converts the mass of that nucleus to the lesser masses of the (usually two) "fission product" nuclei and the mass of two to three neutrons plus about 200 MeV of kinetic energy (energy of motion) of these particles, plus the energy of one or more "gamma rays" (similar to X-rays). Most of this energy is in the form of kinetic energy of the two recoiling intermediate-mass "fission product" nuclei, plus the considerable kinetic energy of the two to three neutrons, gamma rays, and neutrinos.

The energy of the recoiling fission product nuclei is deposited in the fuel as thermal energy within a few millimeters or so of the fission event. The neutron kinetic energy and the gamma energy are deposited in the surrounding fuel, structure, and coolant mostly within 10–20 cm of the fission event, as a thermal source, and the neutrons produce additional energetic gamma rays by capture (n, γ) reactions. Thus, most of the nuclear energy produced at a point in a fission event is quickly converted to a thermal heating source over a 10–20 cm region surrounding the original fission event, although high-energy neutrons and gamma rays distribute some energy further.

This resulting nuclear heating source is peaked in the center of the fuel pins, causing the thermal energy to be conducted outward to the flowing coolant, which carries it to a heat exchanger, where it heats a "secondary system" coolant to produce steam, which in turn drives a turbine generator to produce electricity, as shown in Figure 7.1. This last part of the electricity production process is very similar to the process used to produce electricity from burning coal or gas/oil. So, in a sense, the coal- or gas/oil-burning "firebox" of conventional fossil fuel power plants is just replaced by a nuclear "firebox".

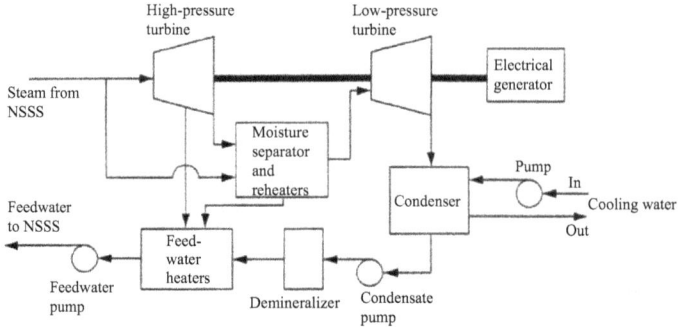

Figure 7.1   Secondary (electrical power producing) side of nuclear reactor power cycle [19]

Chapter 8
Advanced fission reactors

Advanced generation-3 versions of the original PWR, BWR, HTGR, and CANDU pressure tube reactors have been developed, and a series of generation-4 reactors are now under development, including the gas-cooled fast reactor (GFR), the lead-cooled fast reactor (LFR), the molten salt reactor (MSR), the supercritical water reactor (SCWR), the sodium-cooled fast reactor (SFR), and the very high temperature reactor (VHTR) to produce high-temperature process heat.

There has been a recent emphasis on small, modular reactors (SMRs) to provide power for small, isolated communities or to be combined sequentially to reduce the financial burden of building a large power plant all at once.

Chapter 9

The Nuclear Reactor Physics, Nuclear Reactor Engineering, Plasma Physics, and Fusion Technology disciplines

While this book is intended to be self-contained with respect to the science, engineering, and mathematics needed to understand the subject matter at an advanced introductory level, those who might wish to carry out more in-depth investigations/calculations will benefit from some familiarity with the literature of the fields of science and technology involved.

The highly developed body of mathematical physics and supporting nuclear data that is used to describe the neutron distribution in a nuclear fission reactor and its interaction with the reactor materials and dependence on reactor geometry is known as *nuclear reactor physics* [5], and the similarly highly developed body of mathematical engineering and engineering data that describes the thermal-hydraulics, temperature, and stress distributions within and heat removal from a nuclear fission (or fusion) reactor is known as *nuclear reactor engineering* [20], although the latter designation is often used to refer to both the physics and engineering. Good combined treatments of nuclear reactor physics and nuclear engineering at an introductory level [18,19] are also available.

The highly developed body of mathematical physics that describes the fusion plasma behavior [30,32,33] is known as *plasma physics*, and the supporting engineering technology is known as *fusion technology* [29,31].

It is not our purpose here to review the historical or current literature of the fields involved in the study of fusion–fission hybrids, but rather to point out where the interested reader might obtain broader and deeper insight.

Broad introductions to both nuclear reactor physics and engineering are given in [18] and [19], and up-to-date and comprehensive advanced treatments of nuclear reactor physics and nuclear reactor engineering are provided in [5] and [20], respectively.

Chapter 10
Safety of nuclear fission power reactors

In more than 17,000 cumulative reactor-years of nuclear power reactor operation (as of June 2019), in 33 countries, there have been only three major reactor accidents, only one of which caused fatalities. Contrary to the scare propaganda promulgated by various antinuclear groups that were formed following the end of the Vietnam War to continue the profitable "protest movement" business, nuclear fission power has been a very safe industry, with one exception – Chernobyl.

The first "nuclear power accident" occurred at the US Three Mile Island PWR in 1979, where a series of malfunctions led to a loss of reactor coolant which was not evident to the operators from the instrumentation available to them, with the result that the appropriate corrective coolant addition action was not taken and the core subsequently melted down. No operator received an excessive dose and no radiation in excess of safety limits was released from the site. Although the core was destroyed, the radiation containment worked well and no one's physical well-being was affected. Three Mile Island was actually an expensive confirmation of PWR safety.

The Chernobyl accident near Kiev in the USSR in 1986 was not so benign. The nuclear power plant accident was enabled by operators disabling safety systems to perform certain tests, ironically tests intended to enhance the safety of the reactor, and abetted by some design deficiencies of the RBMK-type reactor, the absence of a strong containment building (a standard feature on Western reactors), and a series of operator and management errors. (The Russians do not deal with such matters leniently—the plant director, chief engineer, and deputy chief engineer were subsequently tried and sentenced to Siberian labor camp for 10 years.)

The Chernobyl accident resulted in 31 early fatalities to plant operators and large radiation doses to over 1000 people. Thirty some odd years after the fact, there have been these 31 fatalities to workers present at the accident plus 25 additional fatalities to first-responders who dealt with the immediate aftermath of the accident. However, the medical data to date also indicate about 20,000 thyroid cancer cases (which are not fatal if diagnosed and treated) in children in the region.

A major international program of technical assistance was undertaken by several international nuclear organizations to bring the RBMK and other early Soviet-designed reactors in eastern Europe up to Western standards, and modifications were made to the 11 RBMK reactors operating at the time in the USSR.

In 2011, a magnitude 9 earthquake followed by 7 tsunamis with waves up to 15 m high inundated the Fukushima Nuclear Generation Park in northeastern Japan.

This disconnected the six boiling water reactors (BWRs) from the power grid and flooded the emergency diesel generators, leading to core meltdown, subsequent hydrogen explosions, and the release of radioactivity at levels about 10% of that at Chernobyl. The radiation exposure to the surrounding public was greater than the safe dose level for 42% of these people in the immediate area. The tsunamis caused further enormous damage and 15,894 deaths, not one of which was attributable to the reactors. *Fukushima was a tsunami accident that included nuclear reactors, not a nuclear reactor accident.*

In summary, there have been 56 deaths to workers who dealt with the Chernobyl nuclear accident, and a thousand other people received large radiation doses in that accident, which may have later consequences, plus significant radiation exposure to nearby people in the Fukushima tsunami accident. These accidents are, of course hugely regrettable, and the lessons learned from them have been put to work to reduce the chances of future such radiation exposures from disabled reactors.

However, these accidents also need to be put into perspective with respect to other everyday risks accepted by society. Worldwide, there were 556 deaths due to commercial airline crashes in 2018, and 1.3 million deaths (40,100 in the United States alone) due to automobile accidents in 2017. Thus, 56 deaths due to one of only three nuclear power accidents over 50–60 years is miniscule compared to accidents we routinely accept in our lives, like airliner crashes and expressway wrecks, because the associated benefits are implicitly judged to outweigh the consequences. The nuclear industry itself actually has a very good safety record.

Chapter 11

The nuclear fission power fuel cycle

The nuclear fission fuel cycle is illustrated for present nuclear reactors in Figure 4.1 [6]. First, the mined natural uranium is processed in centrifuges to enrich the $^{235}U_{92}$ concentration of uranium from 0.72% in natural uranium to 4–6%; this enriched fuel is fabricated into metal-clad fuel rods (or pins), which are loaded into nuclear reactors, leaving behind as residue slightly depleted (in $^{235}U_{92}$) uranium and the "tails," consisting basically of $^{238}U_{92}$ and less than 0.2–0.3 wt% $^{235}U_{92}$. The tails are retained by governments for future use as fuel or in other applications where heavy atomic mass material is needed.

These fuel pins are loaded into nuclear reactors to produce power by fissioning some of the original $^{235}U_{92}$ and some of the bred plutonium until there are too few fissionable atoms remaining to be effective, which is 1–2 years, depending upon the enrichment. A 1,000 MWe PWR or BWR with initial fuel enrichment of 4/5/6% in $^{235}U_{92}$ can operate effectively to a fuel burnup of about 40/55/70 GWd/ton. About one-half to two-thirds of this power is produced by fissioning the $^{235}U_{92}$ in the original fuel and the remainder is produced by fissioning the $^{239}Pu_{94}$ and $^{241}Pu_{94}$ produced by neutron capture in $^{238}U_{92}$ (in the second transmutation/decay chain shown in Figure 3.6).

In any natural or industrial process there is always some residue left over. Since the beginning of time, man has just dumped his residue into the rivers, oceans, or air, or buried it in the earth. This worked for a while when there were not so many inhabitants on the earth and they were not so inventive. However, the massive fish kills, rivers full of industrial pollutants that catch on fire, air that is not safe to breathe, and the heating of the earth (as indicated by the melting of glaciers and polar ice caps, the flooding of coastal cities, the increasing number and intensity of powerful storms, etc.) that are upon us indicate that we need to become even cleverer pretty quickly. There is strong evidence (e.g., [2]) that the present "global warming" crisis is due in large part to the dumping of carbon from coal- and oil-fired power plants into the atmosphere, where it forms carbon dioxide, which traps some of the reflected fraction of the incident solar energy within the earth's atmosphere, thereby increasing the temperature of the earth's atmosphere. (Coal- and oil-fired power plants also emit other nasty things that kill trees, fish, people, etc., and corrode metal.)

Since we are advocating the replacement of coal, oil, and other carbon-based fuels for electrical power production with nuclear power plants in order to stop dumping carbon into the earth's atmosphere, it is incumbent upon us to address the question of "what is the residue left over from nuclear fission power plants and their associated production facilities and what should we do with it?"

Nuclear fission power reactors also have a fuel cycle, in which "residues" are created at various stages. However, these residues are not dumped into the environment.

The metal-clad fuel pins are loaded into nuclear reactors and neutron-fissioned to produce power. The length of time that the fuel pins are left in a nuclear reactor is determined by a combination of factors, principal among which are (1) the declining effectiveness of the fuel pin in maintaining the neutron fission chain reaction as the number of fissionable $^{235}U_{92}$ atoms decreases and the number of parasitic neutron capture reactions in the growing number of fission product atoms increases; and (2) "radiation damage"—the general decrease in structural integrity of the fuel pin (swelling, reduction in ductility, reduction in strength, He and H gas accumulation, etc.). After about 18 months of reactor operation (in a typical PWR), "spent" fuel assemblies will be removed and replaced with fresh fuel assemblies. Present practice is to temporarily store the "spent" fuel assemblies in onsite water cooling pools (in order to remove much of the radioactive "decay heat" of the fission products and the transuranics which have been created by neutron capture within the fuel) pending "ultimate disposal."

A method of ultimate disposal of "spent nuclear fuel" (SNF) that has been identified, but not yet widely implemented, by the US and other governments, is "permanent" burial in "high-level waste" geological repositories (HLWRs). Because the "half-lives" (times in which the amount of radioactivity decreases via radioactive decay by a factor of 2) for several of the transuranics produced in SNF (see chapter 6 in [5]) are greater than 100,000 years, the integrity of these HLWRs must be assured for such periods, which complicates their engineering design and greatly increases their cost.

This "spent fuel" contains the intermediate atomic mass fission products (which are strong parasitic neutron absorbers), the remaining fissionable $^{235}U_{92}$ atoms that have not been fissioned or transmuted, and some of the original $^{238}U_{92}$ atoms, some of which have been transmuted into fissionable isotopes ($^{239}Pu_{94}$, $^{241}Pu_{94}$, other higher transuranics). In the plutonium recycle "breeding" fuel cycle the plutonium isotopes are extracted and mixed with uranium isotopes in fuel for subsequent reactors.

Moreover, it has been noted that while the used fuel is not strongly fissionable in the type of thermal reactor (usually a PWR or BWR) neutron spectrum from which it has been removed, it is highly fissionable in a fast reactor (see Figure 6.3). So, it has been suggested that, rather than bury this still-fissionable fuel in a HLWR, it would be better to first further fission it in fast "transmutation" reactors (e.g., [22]). Since fission requires neutrons, which are in short supply in a critical nuclear reactor, a variant on this would be to fission it in a fast transmutation reactor embedded in a large fusion neutron source [11,23,24], *a fusion–fission hybrid* (to which we will return after discussing fusion).

Chapter 12

Scientific and technological basis of nuclear fusion power

The nuclear fusion energy release process is different than the nuclear fission energy release process discussed previously, although the basic process of conversion of mass to energy is the same. As discussed above, the nuclear fission energy release results from the neutron-induced fission of a heavy mass atomic nucleus (e.g., uranium) into a few intermediate mass atomic nuclei and a few basic nuclear particles (e.g., neutrons, protons, alpha particles), with a net reduction in total mass, which is converted to energy via $\Delta E = -\Delta mc^2$ (c is the speed of light, $\Delta m \leq 0$).

On the other hand, the nuclear fusion energy release results from the combination (fusion) of two light mass atomic nuclei to form an unstable compound nucleus that breaks up into other light mass nuclei and nuclear particles which in sum are less massive than the two original atomic nuclei that fused, also resulting in a net reduction in mass that is converted to energy via $\Delta E = -\Delta mc^2$.

Whereas neutron-induced fission takes place in many of the heaviest elements, fusion takes place among many of the light elements. The fusion reactions of primary interest are shown in Table 12.1.

Table 12.1 Fusion reactions of primary interest [28,29]

Reaction	Energy release (MeV)	Threshold energy	
		Kelvin	keV
$D + T \rightarrow {}^4He + n$ (14.1 MeV)	17.6	4.5×10^7	4
$D + D \rightarrow T + p$ (50%)	4.0	4.0×10^8	35
$D + D \rightarrow {}^3He + n$ (2.5 MeV) (50%)	3.25	"	"
$D + {}^3He \rightarrow {}^4He + p$	18.2	3.5×10^8	30

The ion energy dependence of the reaction rate parameters for three fusion reactions with large cross sections at the lowest energies are shown in Figure 12.1.

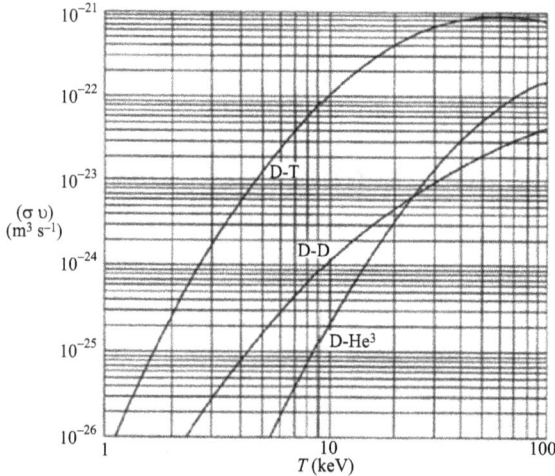

*Figure 12.1   Ion energy dependence of fusion reaction rates of primary interest
[29,30]*

The D–T fusion reaction with the highest probability (cross section) at and just above plasma temperatures that have already been achieved in the laboratory is for the fusing of deuterium (hydrogen with one proton and one neutron in the nucleus) with tritium (hydrogen with one proton and two neutrons in the nucleus) to produce a 3.5-MeV helium ion (alpha particle) and a 14.1-MeV neutron. The charged helium ion (and its energy) is magnetically confined and remains in the plasma to collisionally heat it, while the energetic neutral neutron escapes to the surrounding region. In a *fusion reactor* this surrounding region is a "blanket" material region where the energy of the neutron is deposited and can be removed as heat energy and converted to electricity, and in which the neutron can then be captured in lithium to "breed" tritium for further fusion reactions. In a *fusion neutron source* for a nuclear fission reactor this surrounding region is the nuclear fission reactor plus the tritium breeding blanket.

In order to overcome the strong Coulomb repulsive force acting between two approaching positively charged ions and enable them to approach each other close enough for the strong, but extremely short-range, attractive nuclear force to produce a fusion of the two nuclei into one, the two particles must be approaching each other extremely rapidly, with "thermal" velocities (V_{th}) corresponding to plasma temperatures of at least 4.5×10^7 K for (D,T) fusion, at least 4.0×10^8 K for (D,D) and (D,^{3}He) fusion, and even larger for the fusion of heavier "light" atoms. These thermal velocities correspond to solar temperatures (of course, because fusion is how the sun and other stars produce energy) and the corresponding thermal velocities are about 2×10^6 m/s. Particles moving unimpeded at these speeds would circle the earth at the equator in a few seconds. This is no problem for the sun and other stars, which are much, much larger than the earth, but it is a problem for fusion in a reactor on earth, where particles of this speed incident on a chamber (~1 m dimension) wall would quickly destroy it. So the first issue is how to confine particles moving so fast in a practical confinement chamber with dimensions of meters for practical times of hours, days, weeks.

The most developed solution to the confinement problem is *magnetic confinement*—to use magnetic fields that everywhere remain within the intended confinement volume without intersecting the wall. The force exerted on a moving charged particle by a magnetic field is perpendicular to both the magnetic field direction and the particle velocity direction, which has the effect of making the particle move along and spiral about the magnetic field lines, with a spiral radius $r_1 = mV_{th}/eB$, which works out to be about 0.5–1.0 cm for the light ions in magnetic fields of several tesla, which are of present magnetic confinement fusion interest. So, if a confinement chamber geometry and a magnetic field configuration can be created such that the magnetic field lines stay within the chamber, without intersecting the wall, then the plasma ions and electrons will follow along and spiral about the field lines and remain within the plasma chamber (at least to leading order). The simplest such chamber geometry is the toroid or donut-shaped chamber illustrated in Figure 12.2, which is known in fusion by its Russian acronym *tokamak*.

Electric currents flowing in electromagnets and in the plasma create magnetic fields in the toroidal plasma chamber. If one forms a fist and points his right thumb in the direction of a current, then the magnetic field created by this current is in the direction of the fingers of his right hand. Thus, the current flowing poloidally (around in the short direction) in the large, outermost toroidal field magnets creates a toroidal field around the plasma chamber in the long direction. A toroidal (axial) current flowing in the plasma around the plasma chamber in the long (toroidal) direction creates a poloidal magnetic field encircling the plasma chamber in the short (poloidal) direction. Neither of these magnetic fields intersect the chamber

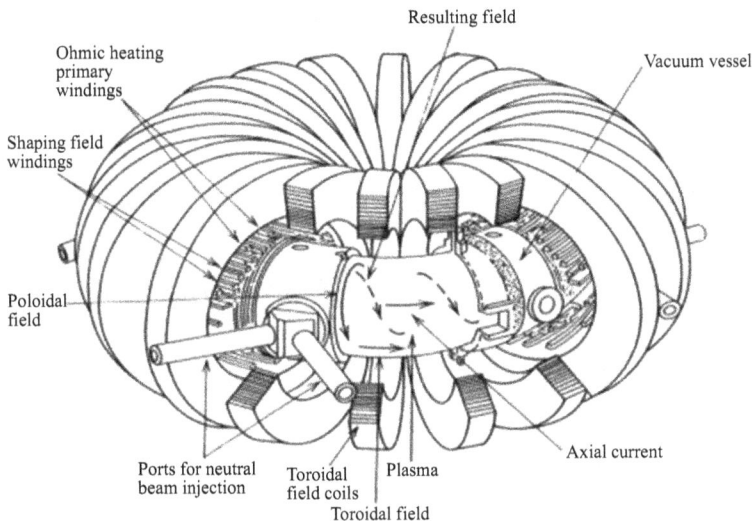

Figure 12.2   Geometry and nomenclature of the tokamak plasma confinement concept [28,29]

wall, so charged particles moving along and spiraling tightly about the combined (toroidal plus poloidal) magnetic field would remain in the plasma chamber, in the absence of other forces. The resulting summed magnetic field spirals about the plasma chamber in both the long toroidal direction and in the short poloidal direction, like the stripes on an old-fashioned barber pole or on a stick of pepper-mint candy. Plasma ions and electrons move along and spiral tightly about the resulting magnetic field and do not strike the chamber wall (at least to leading order), as illustrated in Figure 12.3 (not to scale).

Good introductory discussions of tokamak physics and technology can be found in [29,31], and more advanced discussions of the plasma physics can be found in several textbooks, including [29,31–33].

At the next level of analysis of the order 10^{20} plasma ions and electrons per cubic meter in a fusion plasma, the force balance reveals that an inward (in small radius r) magnetic pressure must balance an outward (in small radius r) ion and electron kinetic pressure. The effectiveness for this purpose of the magnetic pressure can be enhanced by shaping the magnetic field within the plasma, which is done with currents in "field shaping coils." The balance between inward magnetic pressure and outward kinetic plasma pressure is sensitive to the details of the distribution of both and is only stable under certain conditions, imposing magneto-hydrodynamic stabi-lity limits on allowable combinations of plasma and magnetic parameters [29–33].

The plasma must be heated from the room temperature of the injected D (or D + T) gas to the solar temperatures needed for fusion. This is largely done (i) by *neutral beam injection* (the injection of energetic neutral D gas atoms that first have been electrostatically accelerated as ions, then neutralized to enable their passing through the complex magnetic field configuration surrounding the plasma chamber,

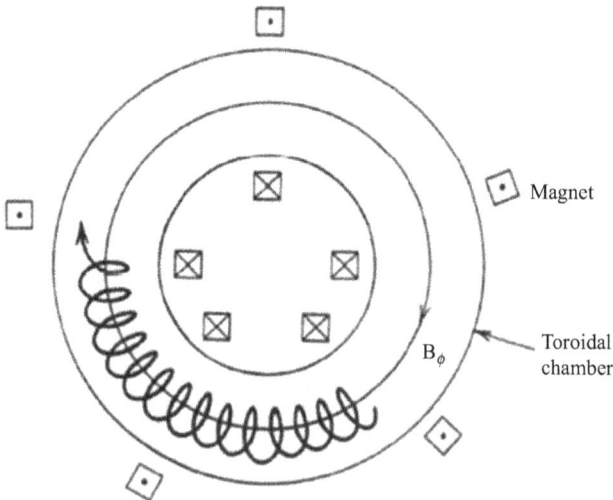

Figure 12.3   Magnetic field and charged particle motion in a tokamak [29] (not to scale)

and finally injected into the plasma, where they are ionized or charge-exchanged and confined magnetically to collisionally transfer their energy to the plasma ions and electrons), and/or (ii) by the *injection of electromagnetic waves* (microwave frequency to radiofrequency), which transfer their energy to heat the plasma electrons or ions (much as a microwave oven heats a pizza).

Lawson criterion
A simple energy balance on the plasma may be used to illustrate a minimum condition of the energy loss rate for which the fusion heating of the plasma can offset the energy loss rate due to transport processes:

$$\frac{1}{4}n^2 \langle \sigma v \rangle_{\text{fus}} E_{\text{fus}} \geq \frac{3nT}{\tau_{\text{E}}} \tag{12.1}$$

where τ_{E} is an energy loss rate time constant from the plasma due to transport processes, and radiation has been temporarily neglected. This can be rewritten as

$$nT\tau_{\text{E}} > \frac{12kE_{\text{fus}}}{\langle \sigma v \rangle_{\text{fus}}/T^2} \simeq 3 \times 10^{21} \text{ keV/ s m}^3 \tag{12.2}$$

Achieved values for this three-factor $nT\tau$ quantity are plotted for various tokamaks since the original Russian T3 in Figure 12.4, demonstrating the several orders of magnitude improvement that has been made (a progress improvement rate better than Moore's law). The ITER experimental reactor that is under construction is expected to perform on the boundary of the region labeled "Reactor Conditions."

When the plasma has been heated to solar energies (>7 keV for a D–T plasma), the fusion rate and fusion heating of the plasma will become significant. When the fusion heating is large enough to offset plasma cooling due to radiation and due to transport of energy to the boundary, then the plasma is

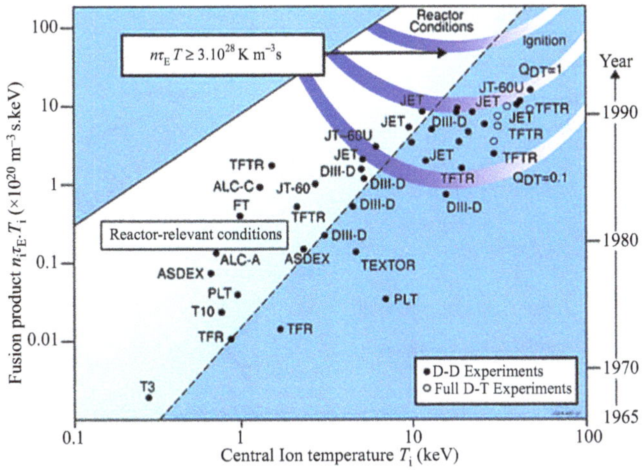

Figure 12.4   Progress of tokamaks in achieving the Lawson criterion [31]

energy self-sufficient without the need for external energy sources (except those *needed for control*). *This condition, known as "ignition," is analogous to a fission reactor being "critical"* and is the holy grail of magnetic fusion research, although not required for a fusion reactor. The ratio Q = (fusion power produced)/(external plasma heating power) is an important performance measure for magnetic fusion research. Reactor design studies indicate that $Q \geq\approx 30$ is needed for economical fusion reactors. A goal of the ITER experimental power reactor project scheduled to begin operation in the early 2020s [32] is $Q \geq\approx 10$. (Plasma energy self-sustainment or *ignition* corresponds to $Q = \infty$.)

In a fusion reactor the 14.1-MeV neutrons and various gamma rays produced in the plasma will pass through the plasma without interaction and emerge into the surrounding blanket region where they will deposit most of their energy as thermal energy via interaction with the blanket material, and the heated blanket coolant will be removed to a heat exchanger to produce steam in the secondary coolant system that will drive a turbine generator to produce electricity, in much the same way as described previously for the fission, neutron, and gamma heating in fission reactors.

However, the energy provided to the plasma by the 3.5-MeV alpha particles (and any other plasma heating energy) will be distributed to the plasma ions and electrons, which will be transported radially outward across magnetic field lines to the plasma edge. If these energetic ions were to come in contact with the chamber wall, then wall atoms would be sputtered into the plasma, where they would cause copious radiation and cool the plasma. To avoid this situation, the outermost magnetic field lines are "diverted" to strike the chamber wall at a location far removed from the plasma, as depicted in Figure 12.5a.

All of the toroidal magnetic field lines inside of the outermost closed field line, designated as the "separatrix," are closed and remain within the plasma chamber and do not intersect the chamber wall. However, all the toroidal magnetic field lines outside the separatrix intersect the chamber wall at the bottom (or top) in

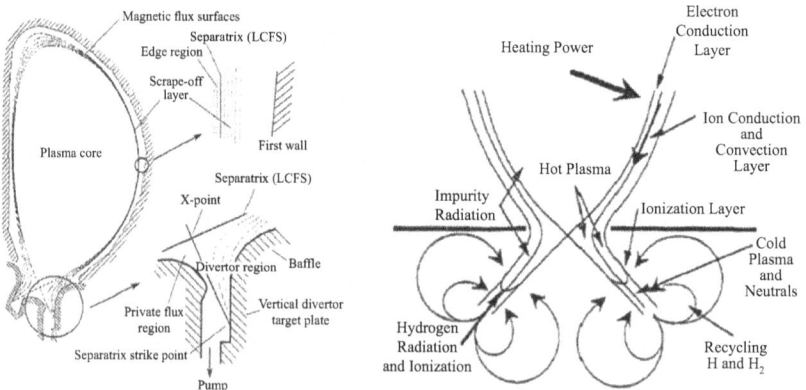

Figure 12.5   (a) Diverted tokamak magnetic field interaction with wall; (b) processes taking place in divertor region [29]

the "divertor" region. So, particles (ions and electrons) spiral about and follow along primarily toroidal field lines in the plasma core which remain within the confinement region, but these particles are transported radially outward (by various processes) across the separatrix onto field lines that do not remain in the confinement volume but run into the chamber wall in the divertor region at the bottom (or top), sputtering wall atoms into the divertor chamber at the bottom (or top) of the plasma chamber. The idea is that these sputtered wall atoms are physically created at far removal from the plasma and would have to "swim upriver" against the outflowing plasma ions to reach the plasma core. The various processes that take place in the divertor region are indicated in Figure 12.5b.

The divertor has proven to be quite effective in maintaining the core plasma relatively free of wall-sputtered impurities. However, the intense particle and energy fluxes incident on the relatively narrow divertor target plate create a challenging heat removal and recovery problem. Solutions are being developed which feature (1) geometric configurations which provide larger incident surface area for the heat and particle fluxes and (2) injection of "impurity" ions to radiate some this energy to other surfaces in order to reduce peak heat loads are both being investigated.

The plasma (D and T) ions incident on the divertor plate are recycled as neutral atoms from the divertor target plates and diffuse back into the plasma chamber to refuel and cool the edge plasma.

A very different scheme for releasing the fusion energy is the so-called inertial confinement in which the fuel (e.g., the D and T atoms) is contained within pellets that are rapidly heated and compressed by laser or particle beams to very high temperatures and densities, where fusion takes place before they can blow apart. Repetitive laser (or heavy ion) compressions are envisioned in such power plants. This process has not yet been as successful as magnetic confinement in producing and sustaining the conditions required for fusion reactors or neutron sources. Other, as yet speculative, schemes for releasing fusion energy without the necessity of achieving solar temperatures are being investigated.

Chapter 13

Nuclear fusion power reactor studies

In 1978–88, the fusion scientists and engineers of the United States, USSR, Europe, and Japan collaborated through the International Atomic Energy Agency's INTOR Workshop [35,36,37] to (1) determine if magnetic confinement tokamak fusion was ready to move forward to the experimental power reactor (EPR) stage; (2) if so, identify a conceptual design of an EPR that combined reactor-relevant physics and technology; and (3) identify and prioritize additional required R&D. Based on the positive results of the INTOR Workshop, Sec. Gorbachev suggested to President Reagan at the 1985 Geneva Summit meeting that the two countries join together to construct and operate the INTOR EPR. This led to the restructuring of the INTOR Workshop into the ITER project [35], today involving the EU, Russia, United States, Japan, China, South Korea, and India, which, after years of negotiations, R&D, and detailed design, is building such a tokamak EPR in France to begin operation in the early 2020s. The design objectives of the superconducting ITER shown in Figure 13.1 are input energy multiplication $Q \geq\ \approx 10$, $P_{fus} = 400\,\mathrm{MW_{th}}$. The toroidal plasma chamber is indicated by the two D-shaped open yellow spaces in the central part of Figure 13.1, with the central solenoidal magnet that induces the plasma current between them. Successful ITER operation will lead to the introduction of fusion power reactors.

The rather large size of ITER is due in part to its role in the fusion development program—it is an experimental reactor in which the fusing plasma will operate and the new technological components will be tested for the first time in the new operating environment of a thermonuclear plasma, and some allowance in size, plasma heating power, etc., has been included in the design to compensate for this exploration of new regimes of operation.

At the present time (2022), ITER is the primary focus of the world's various tokamak fusion research teams (in the United States, Europe, Russia, Japan, China, South Korea, India, and elsewhere). Experimental techniques (1) to mitigate undesirable disruptions, in which the plasma thermal and electromagnetic energy is rapidly transferred to the wall, producing intense heating and large electromagnetic forces; (2) to mitigate or avoid edge-localized modes (ELMs), in which a magnetohydrodynamic instability causes pulses of particles and energy in the edge plasma to be rapidly deposited on the wall or divertor plate; and (3) to optimize plasma density and temperature profiles for maximum performance, etc., are being developed and confirmed on major tokamaks around the world. Aspects of theoretical

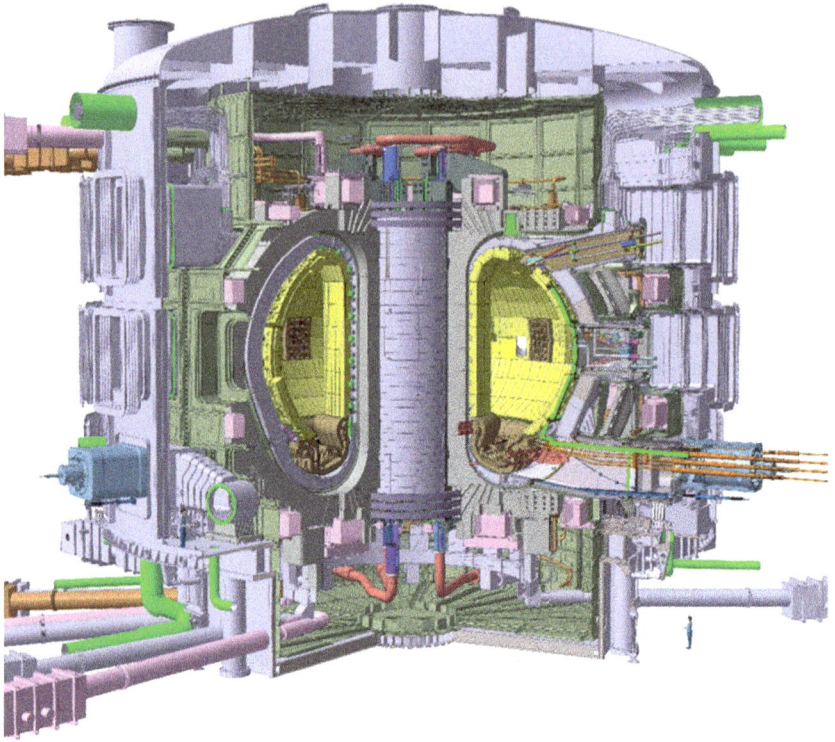

Figure 13.1   ITER tokamak experimental fusion power reactor [35]

and computational analyses of ITER are being tested in present tokamaks, to the extent possible, and intercompared among teams.

Anticipating some improvements in both tokamak plasma physics and fusion reactor technology relative to the ITER design basis allows the design of more compact future fusion reactors, as illustrated in cross section in Figure 13.2 for the design of the 1,000 MWe ARIES-AT reactor design.

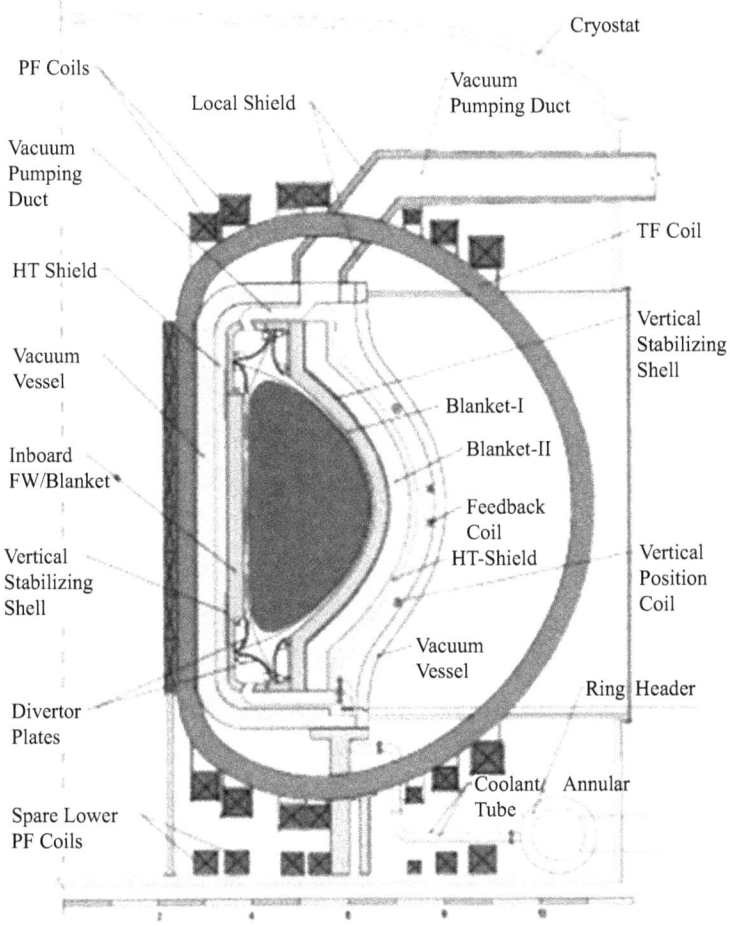

Figure 13.2 The ARIES 1,000 MWe advanced tokamak design [39] (horizontal dimension is about 8 m from centerline on left to outside of vacuum vessel and about 11 m to outboard of toroidal magnet at the right)

Chapter 14

Future improvements in fusion physics and technology

While it is generally believed that ITER will achieve its objectives to explore the joint operation of a thermonuclear plasma and thermonuclear reactor-relevant technologies in a fusion tokamak environment, it is also generally believed that both the plasma operational parameters and the technology must subsequently be improved relative to that available to support the ITER design basis, based in part upon new knowledge from the operation of ITER, but also based in part on new physics and technology insights which will lead to more technically practical and economically competitive fusion reactors in the future.

While anyone working in fusion could draw up a laundry list of advances that would be helpful in achieving an economical fusion reactor, (1) disruption-free and major MHD instability mode-free operation (hence minimization of major pulsed stress and heat loads on the tokamak); (2) higher plasma power density (hence smaller size of everything); (3) more "robust" superconducting magnets that operate at room temperature and are more readily maintainable; and (4) more "neutron radiation-resistant" materials that would not need to be replaced (too often) would place at the top of most such lists. All of these things are being worked on.

Although the tokamak is by far the most advanced magnetic confinement concept in terms of experimental confirmation (see Figure 12.4), there are other confinement concepts that are believed (at least by their proponents) to have various superior long-range reactor prospects. The stellarator concept is similar to the tokamak, except importantly it does not have a plasma current, so no damaging disruptions (i.e., no large pulsed pressure and heat loads on the plasma chamber wall). However, stellarators require a more sophisticated magnetic field coil system, in principle allowing them to operate at steady state. In practice, it has proven extremely difficult to accurately manufacture and install the magnets that precisely produce the more sophisticated magnetic field configuration desired, but there has been progress.

There also have been recent experimental advances in the magnetic tandem mirror, the reversed field pinch, the spherical tokamak, and other less developed plasma confinement concepts; and there are yet other new magnetic confinement concepts only recently proposed or evolved. Experience with all of these concepts is orders of magnitude less than experience with the tokamak, as quantified by the Lawson criterion values shown in Figure 14.1, however, and almost certainly the

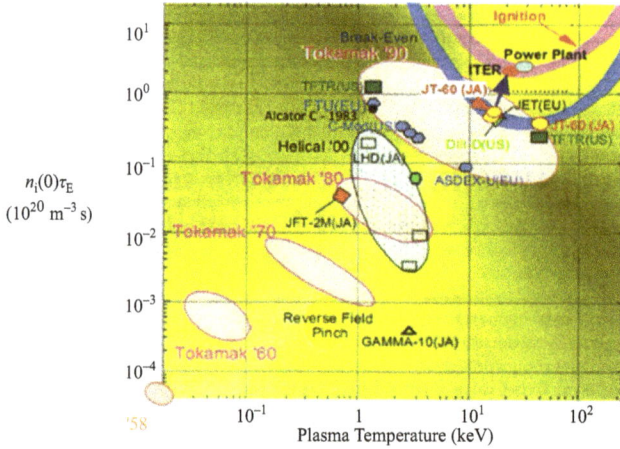

Figure 14.1   Comparison of Lawson $nT\tau_E$ parameter achieved by different magnetic confinement concepts (LHD and Helical are stellarators, GAMMA-10 is a mirror, RFP is a reversed field pinch, and the others are tokamaks) [29]

first fusion power reactors will be based on the tokamak magnetic confinement scheme, and ITER operation will give us a good idea of how well they will work.

Chapter 15
Fusion–fission hybrid reactors

It is clear from the foregoing discussion that a major limitation on the extraction of nuclear energy from uranium (and thorium) fuel ores is the relative scarcity in fission reactors of neutrons, in addition to those necessary to maintain the fission chain reaction, which could be used to transmute the (almost) non-fissionable U238, which constitutes 96% of uranium ore, into fissionable Pu239 and Pu241 or to transmute the non-fissionable Th232 into fissionable U233 or U235. This situation can be remedied by operating fission reactors with a neutron source to provide enough additional neutrons to sustain the neutron fission chain reaction and to capture neutrons in non-fissionable U238 to transmute it into fissionable Pu239 and Pu241 (see the second transmutation chain in Figure 3.6) or capture neutrons in non-fissionable Th232 to transmute it into fissionable U233 and U235 (as shown in the first transmutation chain in Figure 3.6).

Hans Bethe and Andre Sacharov are both credited with suggesting in the middle of the last century that D–T fusion, which is a copious source of neutrons, be used as a neutron source for nuclear fission reactors, which could use some more neutrons in order to access the truly enormous energy content of uranium ore rather than just a small fraction of it. This suggestion has become known as a fusion–fission hybrid (FFH) reactor. Because of the relatively less-developed status of nuclear fusion power and the relative plenitude of uranium, at least for the moment, this suggestion has not been followed up upon until recently. The USSR (now Russia) has had an FFH DEMO reactor as the centerpiece of their fusion development program [41,42] for some time, and China has more recently instituted a broad investigation of FFHs [40]. Less extensive FFH research programs have started in other countries.

The Russian DEMO-FNS is a superconducting tokamak fusion neutron source and, a small subcritical fission core with a minor actinide fuel that generates 200 MW power to maintain the facility. The facility is designed to obtain a steady-state neutron wall loading of 0.2 MW/m^2 and a lifetime neutron fluence on the first wall of 0.2 MW-yr/m^2. The minor and major radii are $a = 1.0$ m and $R = 3.2$ m. The neutral beam (NBI) locations are also shown in Figures 15.1 and 15.2.

The Chinese [40] are investigating a range of different FFH options based on tokamak, spherical torus, and gas dynamic traps, as described in Figure 15.3 and Table 15.1.

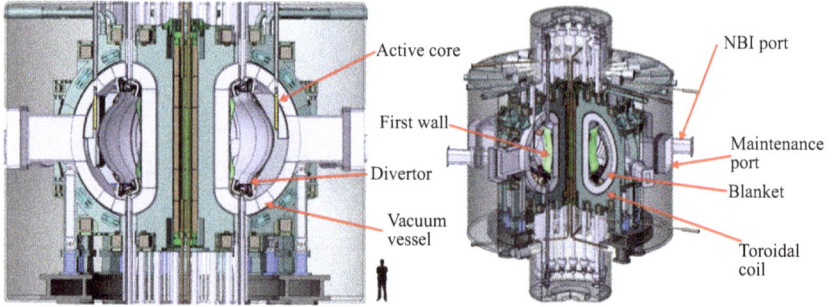

Figure 15.1   A general view of the DEMO-FNS tokamak [42]

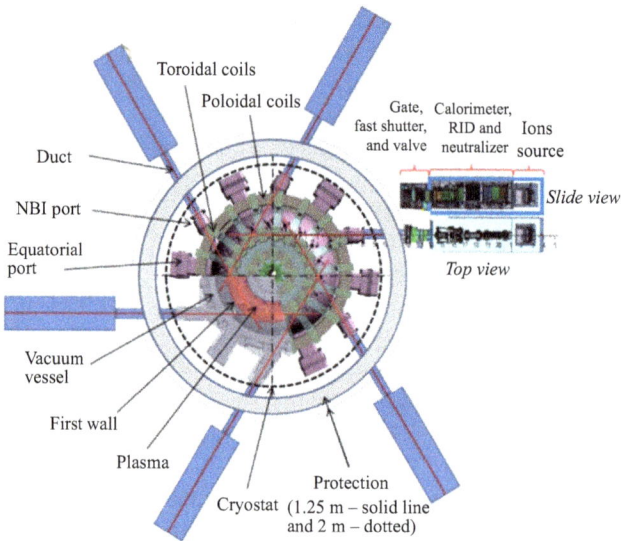

*Figure 15.2   The location of the NBI components on the DEMO-FNS
tokamak [41]*

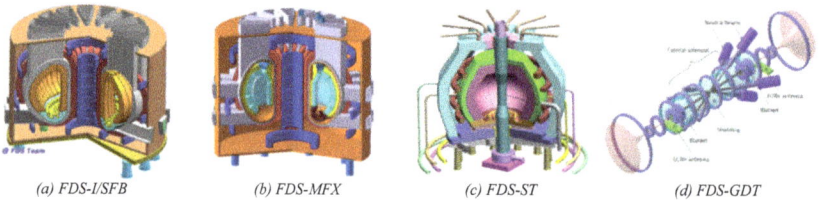

(a) FDS-I/SFB   *(b) FDS-MFX*   *(c) FDS-ST*   *(d) FDS-GDT*

Figure 15.3   Schematic view of FDS series of hybrid reactor concepts [40]

Table 15.1   The main parameters of the Chinese FDS series of FFH reactor concepts [40]

Parameters	FDS-SFB	FDS-MFX	FDS-ST	FDS-GDT
Fusion power (MW)	150	50	100	15
Major radius (m)	4	4	1.4	–
Minor radius (m)	1	1	1.0	–
Neutron wall loading (MW/m^2)	0.49	0.17	1.0	2.0
Fuel	Spent fuel	Depleted/natural/ enriched uranium	Spent fuel	Spent fuel
Coolant	PbLi and helium	PbLi and helium	PbLi and helium	PbLi and helium
Structure material	CLAM	CLAM	CLAM	CLAM

Chapter 16

Principles and technical rationale of the fusion–fission hybrid breeder and burner reactors

The fusion–fission hybrid (FFH) reactor concept is basically a subcritical nuclear fission reactor supported by an additional D–T fusion neutron source to sustain the neutron chain fission reaction for additional energy production, destruction of "nuclear waste," breeding of fissionable material, and/or other neutron applications of the nuclear fission reactor. Such a subcritical FFH reactor would almost certainly be more complex and expensive than an "equivalent" critical fission reactor (CFR). Thus, an FFH reactor would be justified only if its overall benefits were sufficiently greater than those of a CFR for a given nuclear mission. In the technically informed opinion of the author (and others), they would seem to be.

For the purpose of discussion, we postulate that there are two primary *nuclear power missions* envisioned for FFH reactors: (1) to *breed* (create) fissionable nuclear fuel (Pu239 and Pu241 and higher fissionable "minor actinides") from the essentially fertile uranium ore (U238) or to create fissionable (U233) and (U235) from fertile (Th232) and (2) to *burn* (destroy) the long-lived, highly radioactive nuclear waste, the most troublesome part of which consists of long-lived, but fissionable, "minor actinides" that otherwise must be stored in high-level waste repositories for thousands of years, until their radioactivity has decayed away. There are certainly many other missions as may be seen from the references in this book, but we will concentrate on these two principal "burning" and "breeding" missions.

It turns out that most of the longest lived, most highly radioactive "nuclear wastes" are fissionable actinides with very long half-lives, the fissioning of which would not only recover additional energy from the "spent" fuel but also eliminate the longest lived "high-level nuclear waste," which is presently thought to require a massive amount of millennial-secure storage in "high-level waste repositories."

Both efficiently breeding of fissionable Pu239 and Pu241 from non-fissionable U238 and breeding of fissionable U233 and U235 from non-fissionable Th232 require many neutrons in addition to those needed to maintain a critical neutron chain reaction – many more neutrons than are generally available in conventional critical nuclear reactors. Similarly, efficient neutron fissioning of "high-level nuclear waste" requires more neutrons than are available in conventional critical neutron chain fission reactors. Although clever design of CFRs can help the situation, the addition of a source of neutrons to the nuclear reactor would seem to be the most effective solution to this "deficit of neutrons" problem in critical nuclear fission reactors.

While working through the comparisons of FFH and CFR for breeding and for nuclear waste destruction has only begun, two general observations can be made at this time: (1) *Fewer separation steps, hence fewer separations facilities, fewer fuel refabrication facilities, and fewer HLWRs would seem to be needed to achieve the "breeding" or "burning" objectives with FFH than with critical CFR burner reactors; and (2) the accident reactivity safety margin to prompt critical is the delayed neutron fraction in critical reactors, but is the much larger margin of subcriticality for subcritical reactors. This latter means that FFHs can be loaded with 100% transuranic fuel (because the degree of subcriticality can be much larger than the delayed neutron fraction of the fuel), while CFR burner reactors must be loaded with a higher β uranium and transuranic fuel mix to achieve an acceptable reactivity safety margin to prompt critical.* This is a major (and I believe largely unrecognized) advantage of FFHs relative to CFRs. On the other hand, CFRs are less complex than FFHs.

Chapter 17

Illustrative future fusion–fission hybrid reactor designs

There have been a number of scoping "designs," at various levels of detail, of fusion–fission hybrid (FFH) reactors published in the literature over the years. Many of these have been individual design studies, but there have been at least two collected sets of studies of particular reactor types for different FFH missions. We will illustrate the genre (1) with a discussion of the recent substantial Georgia Tech subcritical advanced burner reactor (SABR) (TRU burner) [12,25] and SABrR (Pu breeder) [48] tokamak reactor FFH design studies and (2) with a brief summary of an earlier, but recently updated, comprehensive series [10,13] of magnetic mirror FFH studies from a few years back. The tokamak is the leading fusion reactor concept in terms of achieved performance, level of development, and level of effort world-wide, while the magnetic mirror is probably the geometrically simplest magnetic fusion reactor concept, which encountered particle and energy confinement problems several years ago, but has more recently achieved more encouraging confinement results.

Both the Georgia Tech SABR (TRU Burner [12,44,25]) and SABrR (Pu breeder [48]) designs are based on (a) the Na-cooled fast fission reactor PRISM/IFR [17] fission technology that is being used for the design of the US Versatile Test Reactor (VTR) and on (b) the superconducting tokamak plasma physics and the fusion physics and technology that has been developed for the ITER experimental fusion power reactor [35], which has begun operation in the early 2020s. So, both the fission and fusion physics and technology needed for this tokamak/fast reactor line of FFH development has been developed and will be further developed and prototyped by a world-wide effort over the coming decades. Both the current SABR (burner) and SABrR (breeder) designs produce 3,000 MWth and are based on similar tokamak physics and technology.

The SABR concept, which combines IFR-PRISM fast reactor technology and the ITER tokamak fusion physics and technology in a burner reactor for the transmutation of transuranics, has also been adapted for a subcritical advanced Pu-breeder reactor (SABrR) that produces fissionable Pu from non-fissionable U238. It was found that basically the same fission and fusion technology, geometry, and major parameters used in the SABR burner reactor can be used to achieve a significant fissile breeding rate (fissile breeding ratio FBR = 1.3), while also achieving a tritium breeding ratio TBR > 1.15 (which has been found computationally to be sufficient to achieve tritium self-sufficiency).

The geometrically simpler tandem mirror fusion confinement scheme was initially favored for FFH application on the basis of being geometrically simpler, but the mirror fusion concept encountered confinement difficulties and was abandoned a few years ago in a number of countries, including the United States. However, recent Russian confinement success with the tandem mirror concept has attracted renewed interest to the mirror concept. We will also summarize earlier and recently upgraded work on tandem mirror FFHs, which are also supported by a world-wide program of plasma physics and fusion technology development, albeit not as large as the R&D program supporting the tokamak.

17.1   SABR#1 Na loop-cooled tokamak FFH design with ANL fuel

The first Georgia Tech SABR (SABR#1) tokamak design was configured by locating a toroidal fission core (red) outboard of the toroidal tokamak fusion plasma (yellow) within a toroidal/poloidal magnet system modeled on the ITER [35] magnet system, as indicated in Figure 17.1.

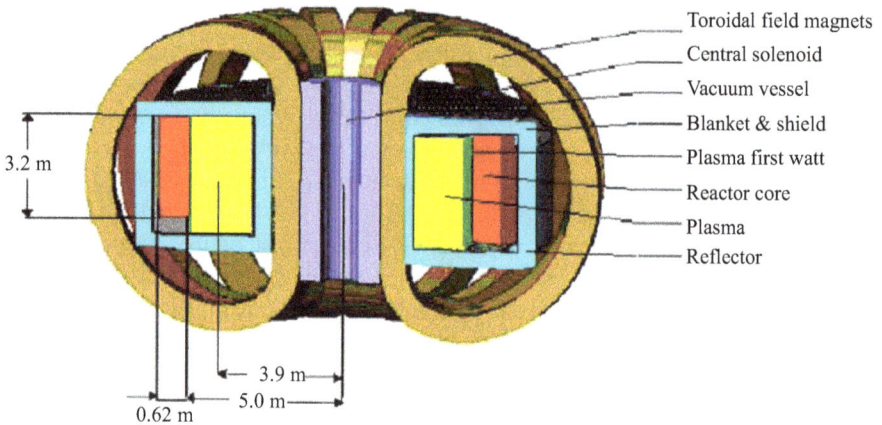

Figure 17.1   3D geometric model SABR#1 [44]

SABR#1 is a Na loop-cooled fast reactor concept based in part on the IFR (PRISM) nuclear fission reactor technology being developed in the United States and in part on the superconducting tokamak fusion technology that has been developed world-wide for the international ITER project [36] being constructed in Europe. A more detailed description of the juxtaposition of the toroidal plasma core surrounded by an essentially toroidal fission core with rectangular plasma chamber cross section, is provided by the r-z neutronics computational model in Figure 17.2 that was used in its analysis.

The neutronics calculations were performed with the ERANOS2.0 code package in P1 transport theory, using the JEFF2.0 neutron cross sections in 33 energy groups. A lattice P1 calculation in 1968 groups was first made for the fission

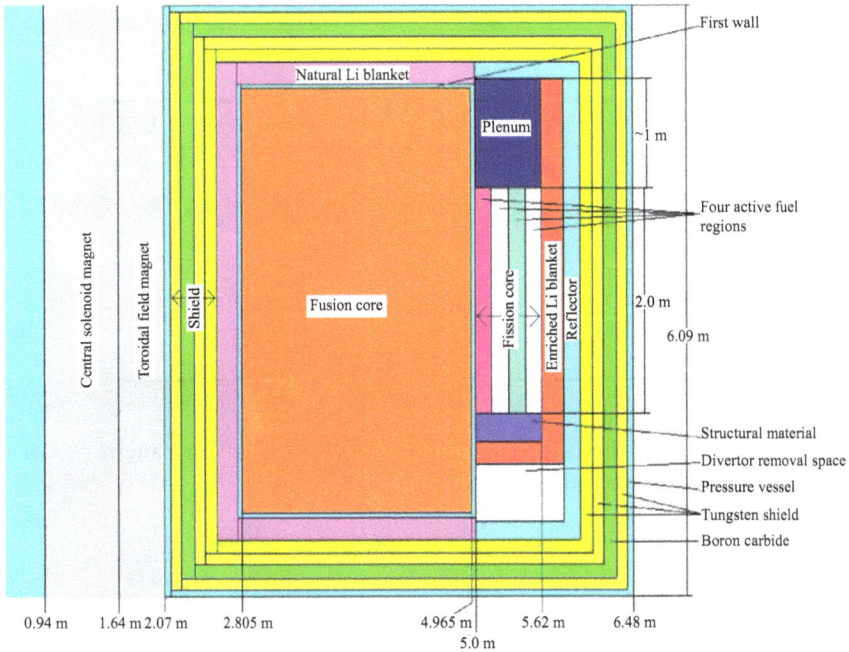

Figure 17.2   SABR#1, r-z neutronics computational model [44]

core (fuel assembly) and collapsed to 33 group cross sections, followed by an S8 33 group lattice transport calculation and homogenization over the fuel assembly. The depletion calculation was then carried out in 233-day time steps.

The fuel-related design parameters for SABR#1 are given in Table 17.1. The "metal" fuel refers to the Argonne TRU fuel of Table 17.2. A metal oxide fuel was also examined, because of European interest.

Table 17.1   Design parameters of metal and oxide fuel pins and assemblies for SABR#1 [44]

Parameter	Metal	Oxide	Parameter	Metal	Oxide
Length of fuel rods (m)	3.2	3.2	Total pins in core	248,778	199,206
Length of active fuel (m)	2	2	Diameter, flat to flat (cm)	15.5	15.5
Length of plenum (m)	I	I	Diameter, point to point (cm)	17.9	17.9
Length of reflector (m)	0.2	0.2	Length of side (cm)	8.95	8.95
Radius of fuel material (mm)	2	3.6	Fuel rod pitch (mm)	9.41	13.63
Thickness of clad (mm)	0.5	0.3	Pitch-to-diameter ratio	1.3	1.56
Thickness of Na gap (mm)	0.83	0.16	Total assemblies	918	918
Thickness of LiNbO$_3$ (mm)	0.3	0.3	Pins per assembly	271	217
Radius of rod with clad (mm)	3.63	4.36	Coolant flow area per assembly (cm^2)	96	108

Table 17.2   The Argonne TRU fuel composition used in
SABR#1 [44]

Isotope	Mass percent at BOL
Neptunium-237	17.0
Plutonium-238	1.4
Plutonium-239	38.8
Plutonium-240	17.3
Plutonium-241	6.5
Plutonium-242	2.6
Americium-241	13.6
Americium-243	2.8

SABR#1 utilizes the out-to-in (right-to-left) fuel shuffling pattern shown in Figure 17.3 to achieve relatively uniform fuel burnup.

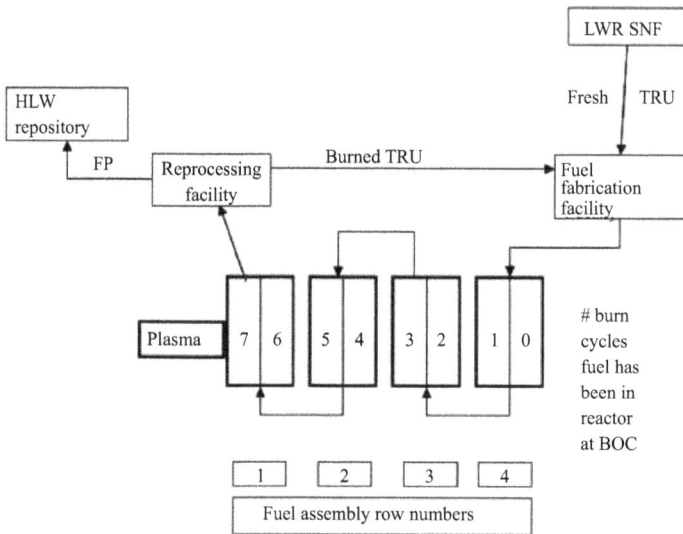

Figure 17.3   The SABR#1 "out-to-in" fuel shuffling pattern (SNF is "spent nuclear fuel," HLW is "high-level waste," BOC is "beginning of cycle) [47]

A major fuel cycle issue is the consequence of clad radiation damage limits on the achievable fractional fuel burnup and fuel residence time. The fuel cycle calculations indicate that the relationship between burnup and radiation damage is about linear from 100 to 300 dpa. Radiation damage limits of 150–200 dpa are anticipated for structural materials presently under development.

A 1,000 MWe PWR produces about 250 kg/year of TRUs, and a SABR#1 operating at 75% availability would destroy (by fission) about 750 kg/year of TRU; hence the 1,000 MWe light water reactor (LWR) support ratio for a SABR#1 is about 3 LWRs (LWR: PWR or BWR), meaning that one SABR#1 would fission all the TRUs produced in three 1,000 MWe LWRs, converting what is now "nuclear waste" that must be disposed of in high-level waste repositories into "nuclear energy" in the process of destroying extremely long half-life nuclear waste.

A 3,000 MWt (thermal) SABR#1 operating at 75% availability would be able to burn all the TRUs discharged annually from 3 LWRs of 1,000 MWe each or would alternatively be able to burn (fission) all of the minor actinides (MAs) and some of the plutonium discharged from 20 to 25 LWRs of 1,000 MWe, while setting aside Pu to start up subsequent fast reactors. Thus, one could envision, e.g., a nuclear fleet with 75% of the energy produced by LWRs and 25% of the energy produced by SABRs that burned the MAs (primarily) and some of the plutonium discharged from the LWRs, while plutonium was being accumulated to start up fast fission reactors. There are other combinations, of course, to increase the nuclear energy extracted from the fissionable fuel while decreasing the mass of long-lived radioactive fissionable actinides that must be buried in high-level waste repositories. The major parameters of SABR#1 are given in Table 17.3.

Table 17.3   Major parameters and materials of SABR#1 [44]

Fission core	
Fission power	3000 MW(thermal)
TRU fuel composition (wt%)	40Zr-10Am-10Np-40Pu
Fuel density	9.595 g/cm^3
Mass of TRU/fuel material	36 t/60 t
Specific power	83.3 kW(thermal)/kg TRU
Maximum k_{eff}	0.95
Major dimensions	$R_{in} = 5$ m, $R_{out} = 5.62$ m, $H_{active} = 2$ m
Fuel pin	Number = 248 778, $D_{fm} = 4.00$ mm, $D_o = 7.26$ mm
Coolant mass flow rate, temperature	$\dot{m} = 8700$ kg/s, $T_{in}/T_{out} = 377/650°C$
Power density, maximum T_{fuel}/T_{clad}	$q''' = 72.5$ MW/m^3, $T_{fm,max} = 715°C$, $T_{clad,max} = 660°C$
Linear fuel pin power	6 kW/m
Clad, wire wrap, and flow tube	ODS ferritic steel, $t = 0.5, 2.2, 2.0$ mm
Fuel/clad, gap, LiNbO$_3$/structure/coolant (vol%)	15/35/14/36
Fuel assembly	Number = 918, hex, $D_{flats} = 15.5$ cm, $D_{side} = 8.95$ cm
Reflector, blanket, and shield	
Reflector/shield materials	ODS steel, boron carbide, tungsten, Na cooled
Tritium breeder	Li$_2$SiO$_4$
Combined thickness	80 cm
Tritium breeding ratio	1.16
Coolant mass flow rate	$\dot{m} = 0.2$ kg/s
Minimum and maximum blanket temperatures	$T_{min} = 450°C/T_{max} = 640°C$
Plasma	
Plasma current	8 to 10.0 MA
Fusion power/neutron source rate	(50 to 500 MW)/(1.8 × 10^{19} to 1.8 × 10^{20} s^{-1})
Fusion gain ($Q_p = P_{fus}/P_{plasma\ heating}$)	180 MW(thermal)/58 MW(thermal) = 3.2

(Continues)

Table 17.3   (Continued)

Superconducting magnets	
Field central solenoid, toroidal field coil, at center of plasma	13.5 T, 11.8 T, 5.9 T
Toroidal field coil magnet dimensions	$w = 5.4$ m, $h = 8.4$ m, $t_{rad} = 43$ cm, $t_{tor} = 36$ cm
Divertor	
Materials	Tungsten, CuCrZr, Na cooled
Heat flux	1 to 8 MW/m^2
Coolant mass flow rate	$\dot{m} = 0.09$ kg/s
First wall	
Materials	Beryllium on ODS, Na cooled
Surface area	223 m^2
Average neutron wall load (14 MeV)	1.0 MW/m^2
Average heat flux (500 MW)	0.25 MW/m^2
Coolant mass flow rate	$\dot{m} = 0.057$ kg/s

17.2   SABR#1 Na loop-cooled tokamak with European fuel type

Somewhat different core designs were developed for three different fuel types of interest internationally. Two different metal fuel types based on metal fuel IFR technology have been developed, one based on the TRU-Zr fuel developed at Argonne (Table 17.2), which is representative of the spent fuel discharged from LWRs, and the other based on the European TRU-MgO metal fuel with reduced Pu (that has been set aside for fueling fast reactors). For initial computational purposes, the SABR#1 fuel pins were based on the composition given in Table 17.1. A separate "European" fuel composition is given in Table 17.4 for Pu saved for fast reactors and MAs to be used in an FFH or other reactor.

Table 17.4   European fuel composition of SABR#1 [24]

Plutonium vector		MA vector	
Isotope	**Mass percent**	**Isotope**	**Mass percent**
Plutonium-238	3.737	Neptunium-237	3.884
Plutonium-239	46.446	Neptunium-239	0.0
Plutonium-240	34.121	Americium-241	75.51
Plutonium-241	3.845	Americium-242m	0.254
Plutonium-242	11.85	Americium-242f	0.000003
Plutonium-243	0.0	Americium-243	16.054
Plutonium-244	0.001	Curium-242	0.0
		Curium-243	0.066
		Curium-244	3.001
		Curium-245	1.139
		Curium-246	0.089
		Curium-247	0.002
		Curium-248	0.0001

Axial and cross-sectional views of the SABR#1 fuel pin are shown in Figures 17.4 and 17.5. The gas plenum is for retention of gaseous fission products.

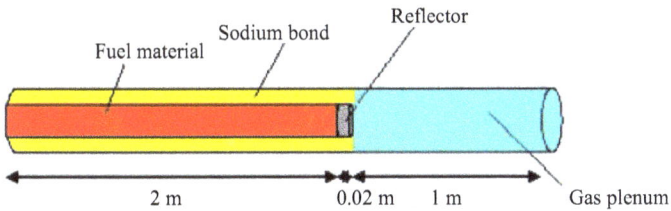

Figure 17.4   Axial view of SABR#1 fuel pin [24]

The cross section of the fuel pin is shown in Figure 17.5, and the basic SABR#1 fuel assembly cross section is illustrated in Figure 17.6.

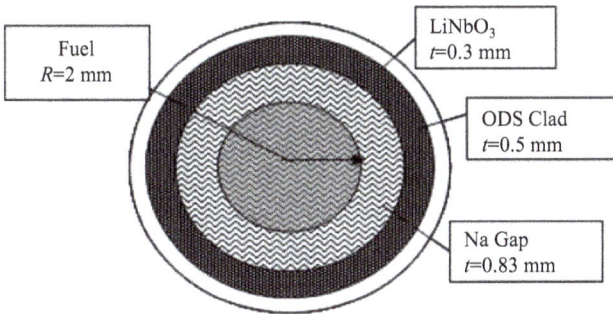

Figure 17.5   Cross-section view of SABR#1 metal-TRU fuel pin [24]

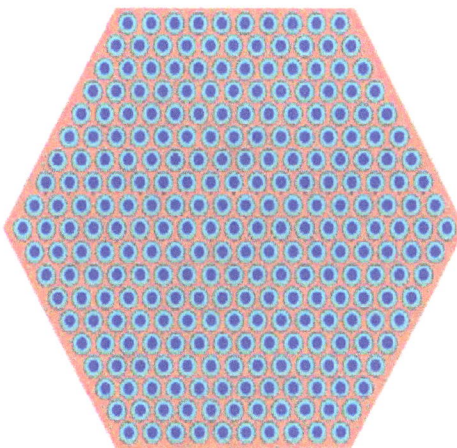

Figure 17.6   Cross section of metallic fuel assembly for SABR#1 (15.5 cm across flats) [24]

If spent fuel is reprocessed to recover plutonium for starting up new reactors, the remaining MAs are still radioactive and fissionable. The European separation of plutonium and "MA-rich" fuels is shown in Table 17.4.

The geometries for the oxide fuel pin and fuel assembly are shown in Figures 17.7 and 17.8.

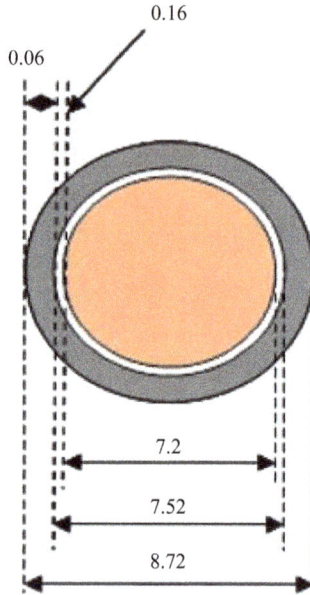

Figure 17.7   Oxide fuel pin [24]

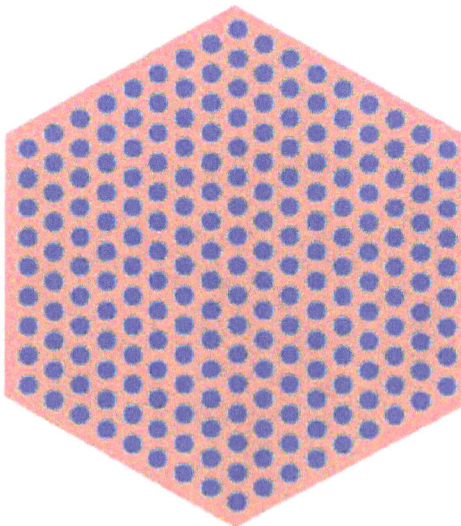

Figure 17.8   Oxide fuel assembly (15.5 cm across flats) [24]

A comparison of the major fuel assembly design parameters for the Argonne metal and EU oxide fuel pins is given in Table 17.5.

Table 17.5   Principal metal and oxide fuel design parameters for SABR#1 [24]

Parameter	Metal	Oxide	Parameter	Metal	Oxide
Length of fuel rods (m)	3.2	3.2	Total pins in core	248,778	199,206
Length of active fuel (m)	2	2	Diameter, flat to flat (cm)	15.5	15.5
Length of plenum (m)	I	1	Diameter, point to point (cm)	17.9	17.9
Length of reflector (m)	0.2	0.2	Length of side (cm)	8.95	8.95
Radius of fuel material (mm)	2	3.6	Fuel rod pitch (mm)	9.41	13.63
Thickness of clad (mm)	0.5	0.3	Pitch-to-diameter ratio	1.3	1.56
Thickness of Na gap (mm)	0.83	0.16	Total assemblies	918	918
Thickness of LiNbO$_3$ (mm)	0.3	0.3	Pins per assembly	271	217
Radius of rod with clad (mm)	3.63	4.36	Coolant flow area per assembly (cm^2)	96	108

17.2.1   SABR#1 fuel shuffling

SABR#1 utilizes the out-to-in fuel shuffling pattern shown in Figure 17.9, in which the fresh fuel is initially loaded into the radially most outboard location (row 4) and shifted sequentially radially inward (row 1). At beginning of life (BOL), fresh fuel

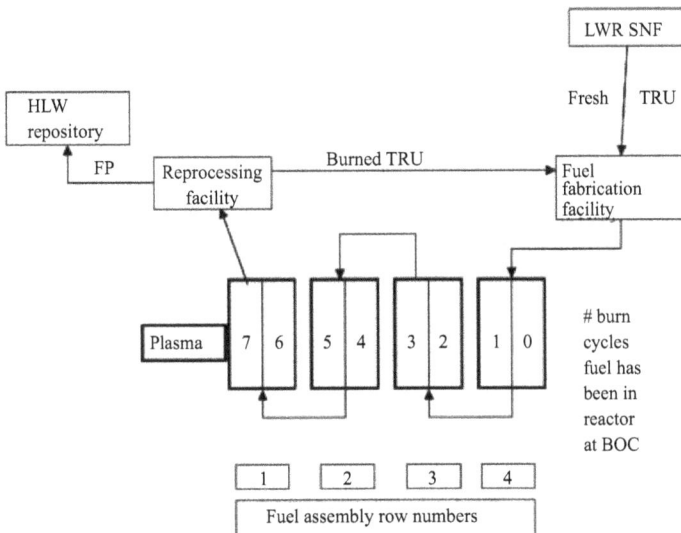

Figure 17.9   The SABR#1 out-to-in fuel shuffling fuel cycle [24]

Table 17.6   European minor actinide-rich fuel composition [24,48]

Plutonium vector Isotope	Mass percent	MA vector Isotope	Mass percent
Plutonium-238	3.737	Neptunium-237	3.884
Plutonium-239	46.446	Neptunium-239	0.0
Plutonium-240	34.121	Americium-241	75.51
Plutonium-241	3.845	Americium-242m	0.254
Plutonium-242	11.85	Americium-242f	0.000003
Plutonium-243	0.0	Americium-243	16.054
Plutonium-244	0.001	Curium-242	0.0
		Curium-243	0.066
		Curium-244	3.001
		Curium-245	1.139
		Curium-246	0.089
		Curium-247	0.002
		Curium-248	0.0001

is loaded in all four annular rows for one burn cycle. After the first burn cycle, fuel in the innermost row (leftmost) is removed for reprocessing, and the fuel in the other seven rows are shifted inward one row, as depicted in Figure 17.9.

The total length of the fuel residence time in the reactor is limited by the clad radiation damage limit. Calculations indicate that the relationship between clad radiation damage and burnup is approximately linear, as shown in Table 17.7.

Table 17.7   Relation between fuel clad radiation damage and burnup for SABR#1 (BOC is beginning of cycle, BOL is beginning of life, EOL is end of life) [24]

Parameter				
Cycle	100 dpa	200 dpa	300 dpa	Once through
Burn cycle length time (days)	350	700	1,000	4,550
Four-batch residence time (yr)	3.83	7.67	10.95	49.8
Fission power [MW (thermal)]	3,000	3,000	3,000	3,000
FIMA (%)	16.7	23.8	31.6	87.2
Region power peaking BOC/EOC	1.7/1.8	1.8/2.0	1.8/2.0	2.0/2.1
BOL P_{fus} (MW)	73	73	73	73
BOC P_{fus} (MW)	155	240	286	1,012
EOC P_{fus} (MW)	218	370	461	1,602
BOL k_{eff}	0.972	0.972	0.972	0.972
BOC k_{eff}	0.940	0.894	0.887	0.784
EOC k_{eff}	0.916	0.868	0.834	0.581
TRUs burned/yr (kg)	1,073	1,064	909	545
Support ratio (75%)	2.9	3.2	3.6	2.2
Clad damage (dpa)	97	214	294	1,537

The TRU burnup rate depends on the fission rate, of course, and the fission rate decreases with fuel burnup, so the fusion rate is increased with TRU burnup to compensate for the reactivity decrease and maintain a constant total power level.

Since a 1,000 MWe LWR produces about 250 kg of TRU/year, a support ratio can be defined as the ratio of the yearly SABR#1 TRU fission destruction rate to the yearly LWR TRU production rate, which is about 3. In other words, a 1,000 MWe SABR#1 could burn about all the TRU produced by three 1,000 MWe LWRs in a given time of operation, thereby both increasing the nuclear energy produced from that fuel by about 33% and reducing substantially the amount of material that must be buried in a high-level waste repository.

17.3 SABR#2 design

The principal difference between the SABR#2 and SABR#1 reactor designs is the change from Na loop cooling in SABR#1 to Na pool cooling in SABR#2 designs in order to investigate and possibly take advantage of passive safety features of Na pool cooling demonstrated for fission reactors in the EBR-II program, although a number of other changes were also evolved.

The basic SABR#2 geometry for both the "burner" and "breeder" reactors is shown in Figure 17.10, where the fissionable fuel is placed in 10 (red) Na pools just outboard of the toroidal plasma chamber.

The fuel is contained in fuel assemblies within 10 separate modular Na pools, as shown in Figures 17.10 and 17.11.

The sodium pool engineering parameters are given in Table 17.8, and the fuel is based on the Argonne fuel composition of Table 17.4.

The layout of the modular sodium pools and the scheme for their removal and replacement is indicated in Figure 17.12.

The ten numbered wedge-shaped sodium pools, the two heating and current drive (HCD) "pools" and the four tritium breeding blankets (TB5–TB8) form a wedge-shaped (in cross section) torus just outboard of the toroidal plasma chamber. Sodium pool removal ports are located between the TF coils at the locations of pools #1 and #6. Refueling and maintenance would be performed by first removing Na pools #1 and # 6 and transporting them to refueling/maintenance rooms, then rotating Na pool #2 to the vacated position of Na pool #1 and rotating Na pool #7 to the vacated position of Na pool #6, and removing both of them separately to refueling/maintenance rooms, then removing the next two pools, etc. Reloading the modular pools would be in the reverse order of the removal process [27].

The maximum duration of the rotation and removal phases (for pools #5 and #10) is estimated to be ~1 h. The maximum decay heat rate that could be tolerated without any heat sink for 1 h without clad damage would be 3.18 MW (~1% of full power). We estimate that reduction of reactor power to 1% full power occurs in ~10,000 s, so it should be safe to disconnect the secondary cooling system by ~3 h after shutdown. Hence, reduction of decay heat to tolerable levels should not cause any significant delay in the refueling or maintenance process.

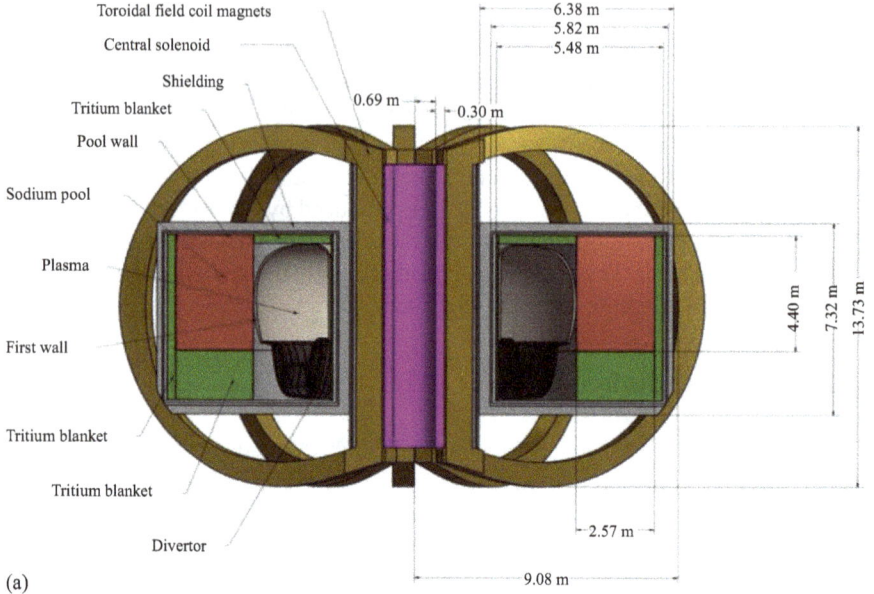

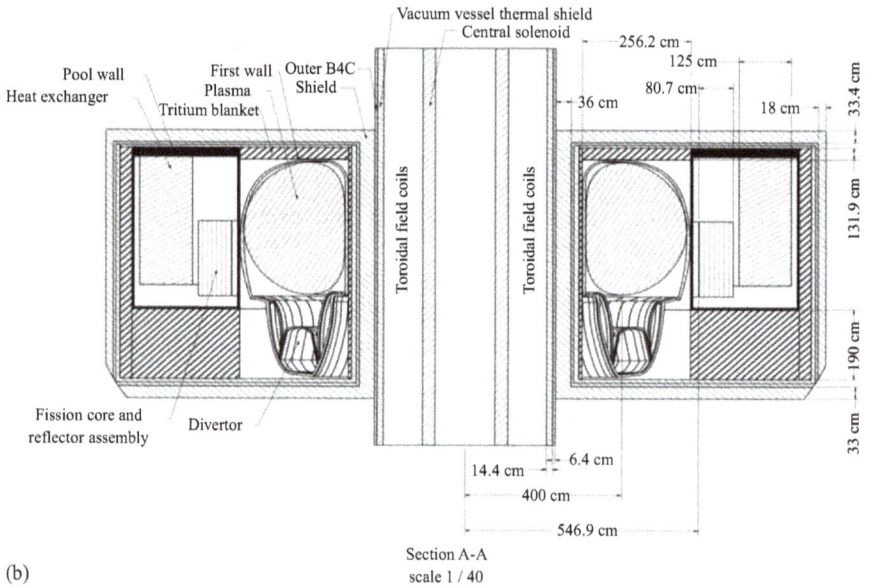

Figure 17.10   The SABR#2 (subcritical advanced breeder/burner) sodium pool fusion–fission hybrid reactor concept [26]

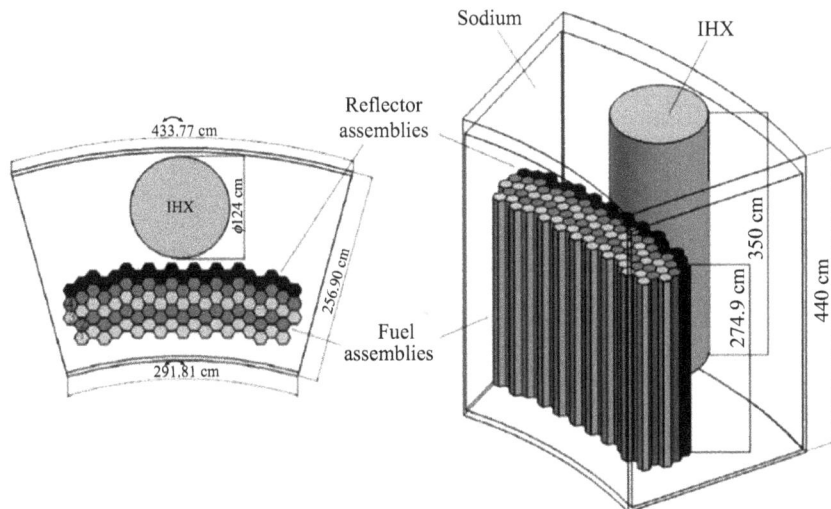

Figure 17.11   A SABR#2 modular sodium pool with fission fuel assemblies and intermediate heat exchanger (IHX) [26]

Table 17.8   Modular sodium pool parameters for SABR#2 spent fuel burner reactor [26]

Number of modular pools	10
Mass of fuel per pool	1,510.4 kg
Mass of Na per pool	22,067 kg
Power per pool	300 MWth
Power peaking	1.27
Mass flow rate per pool	1,669 kg/s
Number of pumps per pool	2
Pumping power per pool (EM pumps)	20 MW
Core inlet/outlet temperatures	628 K/769 K
Fuel max temp/max allowable temp	1,014 K/1,200 K
Clad max temp/max allowable temp	814 K/973 K
Coolant max temp/max allowable temp	787 K/1,156 K

17.4   SABR#3 design

SABR#3 [49] is the same physical reactor as described for SABR#2 and depicted in Figures 17.10–17.12. However, whereas the ten Na pools were assumed to have identical properties (neutron densities, temperatures, etc.), i.e., were modeled with a point kinetics reactor dynamics model for safety analysis, in SABR#2, separate but coupled neutron densities, temperatures, stresses, etc., were calculated for each of the ten different sodium pools in SABR#3, i.e., SABR#3 was modeled with a

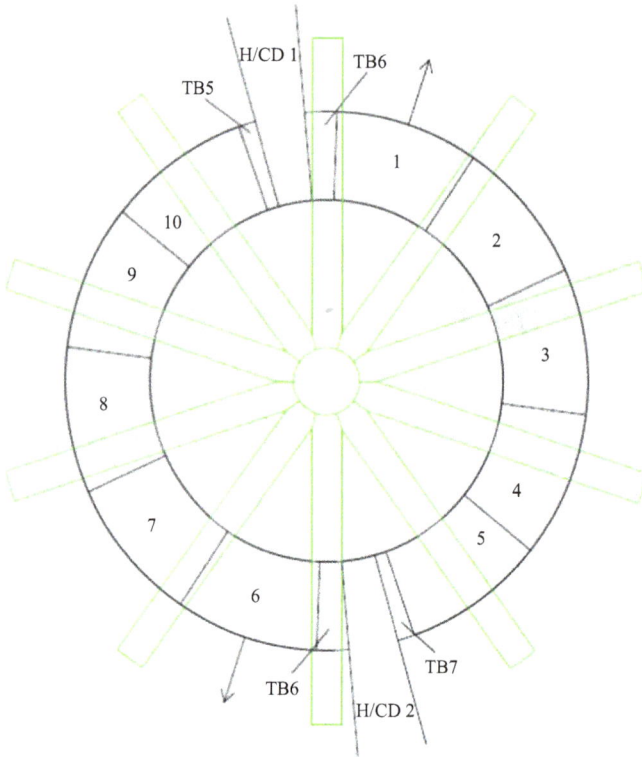

Figure 17.12   Layout of modular sodium pools and replacement scheme in SABR#2 [26]

"ten-node coupled core space-time dynamics model" [49], with the neutronics coupling among cores being calculated with Monte Carlo, in order to evaluate the overall reactivity effects. The reactivity perturbations taken into account in evaluating reactivity effects were core grid plate expansion, fuel axial expansion, fuel bowing, thermal Doppler broadening, and sodium expansion or voiding, all of which conceivably could be different in the different nodes under different off-normal conditions. Our ten-node calculations did not indicate any tendency for dynamic flux tilting oscillations, nor is there any reason to expect any intrinsic flux tilting oscillations in a fast reactor (Figure 17.13).

17.4.1   Thermal property data and empirical correlations

The thermal-hydraulics code COMSOL is provided the material properties for the fuel, cladding, and sodium as well as the empirical correlations for the friction factor and Nusselt number. It is assumed that the properties of the fuel, cladding, and heat

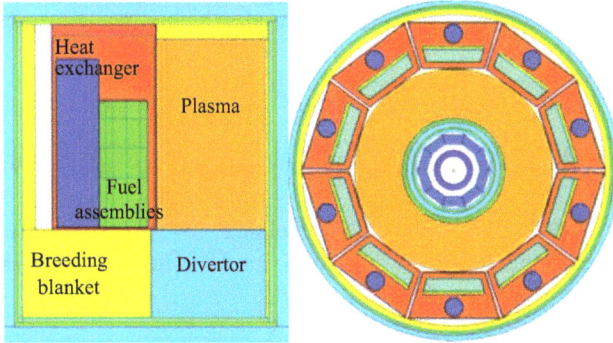

Figure 17.13   Side and top-down views of SABR#3 3D MCNP multinodal model [49]

exchanger tube are determined by the same formulas in the different cores, but using the local nodal temperatures. The properties are given in Table III [12,13].

17.4.2   Design basis accidents

It is standard reactor design practice to evaluate the dynamic response of the reactor design to a set of "design basis accidents" such as loss of coolant accident (LOCA), loss of flow accident (LOFA), loss of power accident (LOPA), etc. This involves a dynamic calculation of the coupled dynamics of the neutrons and of the changing materials, coolant and flow properties resulting in the reactor.

The ANL sodium property correlations are used, and they are given by

$$k = 124.67 - 0.11381 \quad T + 5.5226\mathrm{e}5 \quad T^2 - 1.1842\mathrm{e}8 \quad T^3, 371 \le T \le 1,500,$$

$$(17.1)$$

$$\ln(\mu) = -6.4406 - 0.3958\ln(T) + \frac{556.835}{T}, 370 \le T \le 2,500, \qquad (17.2)$$

$$\rho = 219 + 275.32\left(1 - \frac{T}{2503.7}\right) + 511.58\left(1 - \frac{T}{2503.7}\right)^{0.5}, \ 371 \le T \le 2,503,$$

$$(17.3)$$

and

$$c_{\mathrm{p}} = 1.6582 - 8.4790\mathrm{e}4T + 4.4541\mathrm{e}7 \quad T^2 - 2,992.6 \quad T^{-2}, 370 \le T \le 1,900$$

$$(17.4)$$

For the friction factor, the Zigrang–Sylvester approximation of the Colebrook–White correlation [15] is used,

$$\frac{1}{\sqrt{f}} = -2 \quad \log\left(\frac{\varepsilon}{3.7D} + \frac{2.51}{\mathrm{Re}}\left(1.14 - 2\log\left(\frac{\varepsilon}{D} + \frac{21.25}{\mathrm{Re}^{0.9}}\right)\right)\right) \qquad (17.5)$$

For the Nusselt number, Eq. (17.6) is used, which is recommended by Seban and Shimazaki [16]:

$$Nu = 5.0 + 0.025(RePr)^{0.8}, Re \geq 100 \tag{17.6}$$

Thermal conductivity = 10 W/mK for the TRU fuel and 30 W/mK for the ODS clad;

Density = 3,861 kg/m^3 for the TRU fuel and 7,692 kg/m^3 for the ODS clad;

Specific heat capacity = 738.15 J/kg-K for TRU fuel and 630 J/kg-K for the ODS clad.

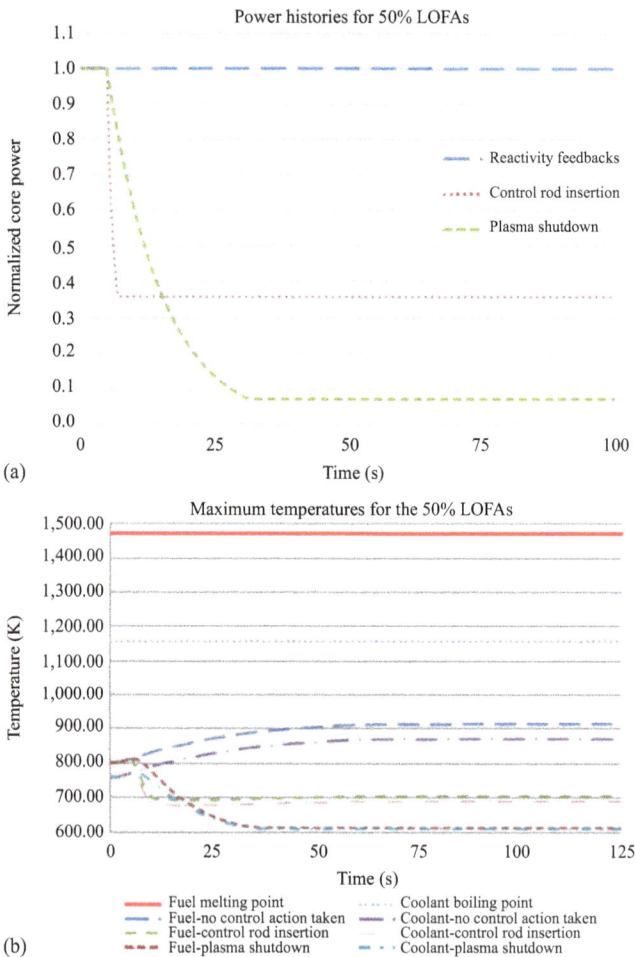

(a)

(b)

Figure 17.14   (a) Power histories for 50% LOFAs and (b) maximum temperatures for the 50% LOFAs [49]

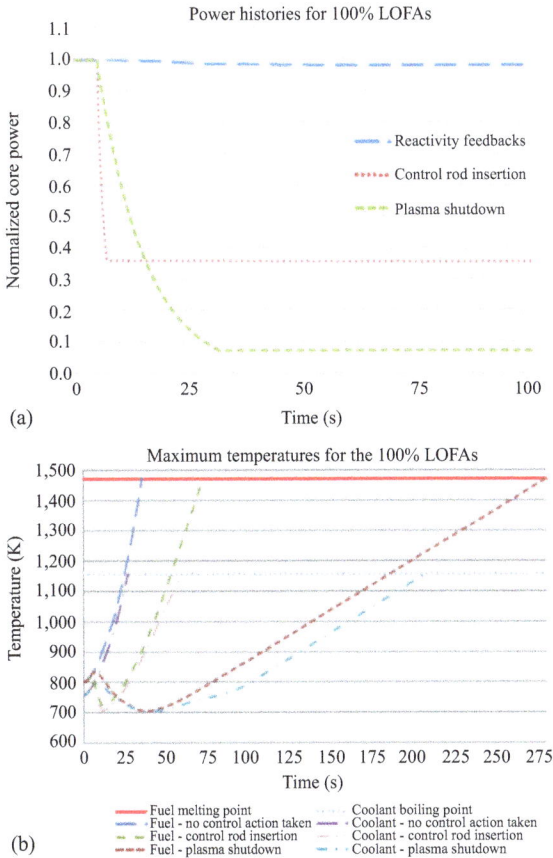

Figure 17.15   *(a) Power histories for the 100% LOFAs and (b) maximum temperatures for the 100% LOFAs [49]*

Dynamic safety analyses of representative malfunctions (loss of coolant, loss of coolant flow, etc.) were simulated. The resulting changes in reactor properties calculated for the failure of 50% and for 100% of the coolant pumps, resulting in a 50% LOFA and a 100% LOFA, respectively, are shown in Figures 17.14. In both composite figures, the first figure is the resulting power level when the indicated control action (none—intrinsic reactivity feedback only; or control rod insertion; or fusion neutron source shut-off) is taken. The second figure shows the temperatures achieved relative to the coolant boiling temperature and the fuel melting temperature.) *Shutting off the fusion neutron source is a very effective control mechanism that can be accomplished in a FFH by opening an electrical switch.* The implication of these and other calculations is that a SABR-type FFH would survive a 50% LOFA, but not a 100% LOFA, without additional design precautions. This has design implications, i.e., different pumps on different electrical systems.

Clearly: (1) shutting down the fusion neutron source (which can be done by flipping an electric switch) is an effective reactor power shutdown mechanism, actually a more effective shutdown mechanism than the insertion of control rods and extending the onset of fuel melting from about 50 to 280 s in this case and (2) the SABR#3 design can survive the accidental loss of up to half of the coolant pumps without any maximum temperatures being exceeded, but the SABR#3 design cannot survive complete loss of coolant pumps without fuel melting and coolant boiling, so additional safety precautions are necessary.

The implication of these loss of coolant flow accident simulations is that the heat removal system on SABR#3 (which is really the Na pool SABR#2 but modeled with a more detailed nodal space–time dynamics calculation model) is sufficiently robust to survive an accident (the 50% LOFA) in which half of the coolant pumps fail, without fuel melting or Na coolant boiling. This clearly indicates that obvious design features like putting different pumps on different power supplies should be observed in future more detailed designs.

17.5   Tandem mirror FFH designs [10,13]

Mirror plasma confinement devices have a number of attractive features as future fusion devices and as FFHs: they have simple linear geometry to readily enable construction and maintenance, are inherently steady state, operate at high beta, have no externally driven currents, and have natural divertors to handle heat loads external to the magnet system, which latter also reduces first wall heat loads. However, the mirror experimental program encountered plasma confinement problems several years ago, and a number of national mirror programs (including the US program) were shut down. Over the past decades, largely after the termination of the mirror program in the United States, several techniques have been suggested and, in some cases, tested experimentally, for making mirrors stable in axisymmetric geometry. The confidence in the practicality of axisymmetric MHD-stable mirrors was increased significantly after a set of experiments conducted in 2005–10 on the upgraded axisymmetric gas dynamic trap (GDT) mirror machine at Novosibirsk, which routinely operates at a plasma beta equal to 0.6 and average ion energy of a few keV, with the plasma axial losses being in good agreement with the classical predictions.

The GDT's important feature is being fully axisymmetric and, at the same time MHD stable [13]. A significant role in making this device MHD stable is played by the outflowing plasma, which, on the one hand, provides a favorable contribution to the stability integral and, on the other hand, provides an electric contact with the conducting end wall [13]. Applying a potential to the segmented limiter transfers to the confinement zone along the field lines a radial potential that may further improve stability. This technique can be used in a fusion neutron source for materials and subcomponent testing with no (or with a minor) extrapolation of the plasma parameters from the existing experiment [13], which will operate at plasma Q of order of a few percent [13].

The attractive features of mirrors are tremendously amplified in the case of axial symmetry. In particular, neoclassical and resonant transport are completely eliminated; engineering simplicity and general flexibility of the device increase significantly; much higher magnetic fields become available for mirror throats, etc. Axisymmetry is thus a game-changer in mirror systems [13].

In order to have a meaningful power balance in such a system, the overall fusion driver has to have a much higher value of Q than the neutron source [13]. A physics background for this more challenging application has been identified and plausible stabilization techniques have been identified and other plasma physics issues affecting the driver performance have been analyzed. The result was a simple, single-cell mirror device with large expansion tanks at the ends indicated in Figure 17.16.

Rather than developing a specific design for a fission reactor coupled to a tandem mirror fusion driver, it was shown in [13] that the GDT is compatible with a broad variety of fission reactors and can perform any of the aforementioned functions. The requirements for the main systems of the facility are discussed in [13]: neutral beam injection system, gas feed and vacuum systems, magnetic system, tritium breeding, and, of course, "blanket" (fission core) and shield. Areas were identified where the required technologies and components are available today and other areas where some further development is needed. The main conclusion was that the hybrid driver in the form of axisymmetric mirrors can be built based mostly on either existing technologies or technologies that will be needed in any of the fusion energy systems (e.g., tritium breeding and neutral beam injection). Further, the axisymmetric mirror hybrid can accommodate fission reactor components designed for other confinement systems.

A schematic of the *tandem mirror* system is presented in [13]. Neutral beams are injected normally to the magnetic axis near the ends of the confinement region where the magnetic field is approximately two times higher than in the uniform regions of the plasma. The maximum magnetic field (in the mirror throat) will be three to four times higher than that at the injection point, which results in the injected ions being well confined. In the uniform section the ions will have a "sloshing" distribution, with an average pitch angle of 45°. Such a

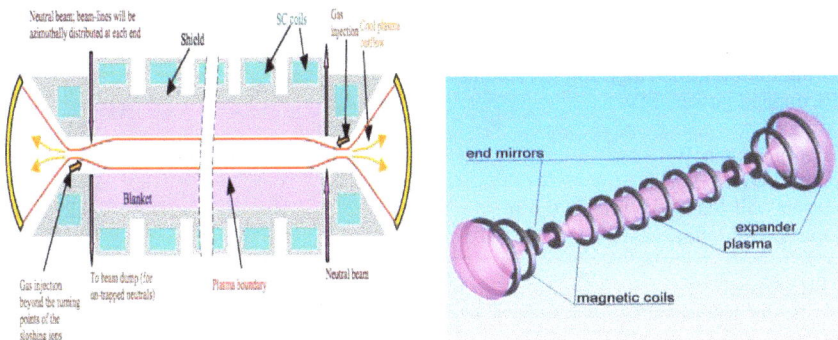

Figure 17.16   The tandem mirror FFH concept ("blanket" indicates the fission core) [13]

distribution has been proven to possess good microstability. The sloshing distribution is compatible only with relatively cold electrons, so that the slowing-down time is shorter than the ion-scattering time. In order to hold the electron temperature low, at the level of 3 keV, the injection of cold atomic streams is envisioned in the zone between the mirrors and the ion turning points. The distance to the ion turning point has to be large-enough to minimize penetration of atoms to the zone with significant hot ion population, in order to minimize charge exchange losses.

Because of their relatively simple solenoidal geometry, the early magnetic mirror plasma confinement concept was initially favored (relative to the toroidal tokamak or the more geometrically complicated toroidal stellarator) for an FFH by some, and a large study of the technical and economic feasibility of producing fissile fuel primarily in tandem mirrors was undertaken in the United States [10] several years ago (Table 17.9).

Since the toroidal geometry of the tokamak is essentially the linear geometry of the tandem mirror "bent" into a torus, a fission reactor designed for either one can conceptually be easily adapted for the other, as illustrated in Figure 17.17.

Table 17.9   Characteristic parameters of a mirror fusion driver for an FFH [10]

Plasma radius[a], m	0.5
Mirror-to-mirror length, m	40
Length of a reacting plasma[2], m	35
Volume of a reacting plasma[2], m^3	25
Plasma surface area[b], m^2	100
Injected ion energy[c], keV	80
Average ion energy[c], keV	40
Average ion density, m^{-3}	10^{20}
Electron temperature, keV	3
Peak ion density, m^{-3}	1.3×10^{20}
Z_{eff}[d]	1.2
Magnetic field, T	2.5
Mirror field, T	15
Volume-averaged beta	0.25
s = plasma radius/average ion gyroradius	30
NBI trapped power, MW	65
Plasma Q	0.7
Fusion power, MW	45
Neutron power, MW	36
Neutron wall load, MW/m^2 @ 0.6 m	0.27
Power to end tanks, MW	75

[a]In the midplane.
[b]Between the turning points of the sloshing ions.
[c]Ignoring ½ and 1/3 energies.
[d]Based on the experience with large-scale mirror facilities and composition of the injected particle beams [14].

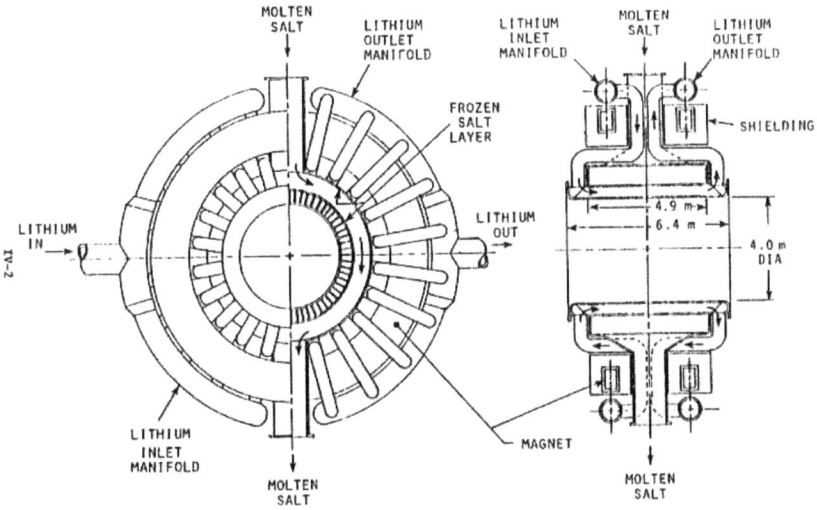

Figure 17.17   Comparable tokamak (left) and tandem mirror (right) FFHs [13]

Table 17.10   Minimum required plasma Q for various versions of the tandem mirror fusion–fission hybrid for the recirculating power fraction = 0.2, $P_{fission}$ = 3,000 MW [13]

Actinide burner			
Blanket multiplication, M	**Minimum Q required**	**P_{fusion}, MW**	**Comments**
Transuranics, $M = 19$	1	200	Solid fuel, engineered or active safety
Minor actinides, $M = 38$–150	0.1–0.5, 0.2 av.	25–100, 50 av.	"
Transuranics, molten salt, $M = 13$	1.5	280	Passive safety
Fuel producer			
Fission suppressed, $M = 2.1$, ^{233}U	8	1,600	Passive safety
Fast fission, $M = 10$, ^{239}Pu	2	370	Engineered safety
Power producer			
$M = 10$	2	370	Molten salt passive safety solid fuel engineered safety
Pure fusion			
$M = 1.34$	11	2,300	Passive safety

Based upon recent Russian advances in mirror fusion physics, the magnetic mirror fusion physics concept has been revived. An upgrade of an old US design concept is shown in Table 17.11.

Table 17.11   Asymmetric tandem mirror fusion–fission physics parameters [13]

Mirror-to-mirror length	$L = 35$ m
Magnetic field in the solenoid	$B_0 = 2.5$ T
Magnetic field in the NBI area	$B_{inj} = 5$ T
Plasma density in solenoid	$n_0 = 10^{14}$ cm^{-3}
NBI power	$P_{inj} = 100$ MW
Plasma beta	$\beta = 0.25$
Plasma radius in the solenoid	$a = 0.5$ m
Magnetic field in the mirror	$B_{mir} = 15$ T
Magnetic field at the end walls	$B_W = 0.04$ T
Injection energy	$W_{inj} = 80$ keV
Electron temperature	$T_e = 3$ keV
Plasma Q	$Q = 0.6$

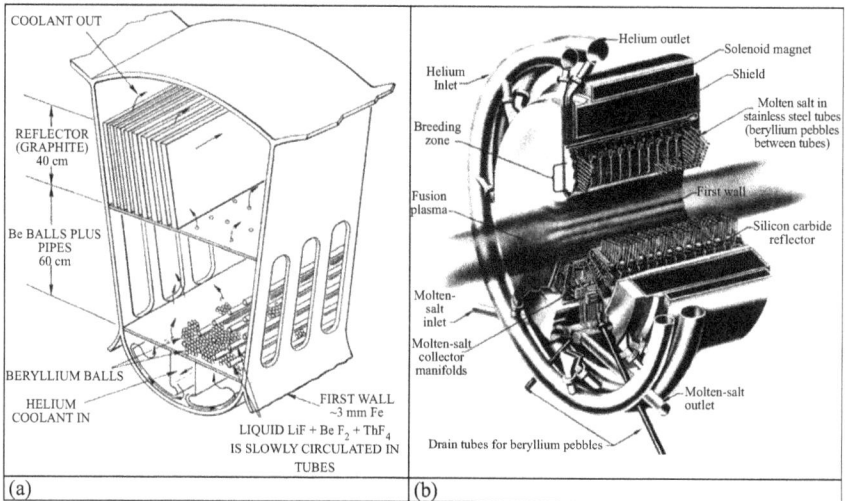

Figure 17.18   Fission module designed both for a tandem mirror (a) and tokamak (b) with pebbles and helium cooling, adapted to mirror geometry making an integrated package of first wall, blanket, shield and, solenoidal magnet [13]

An important performance parameter for any magnetic fusion device is the ratio of Q = power produced/power input. The expected values of Q in tandem mirrors is shown in Figure 17.19.

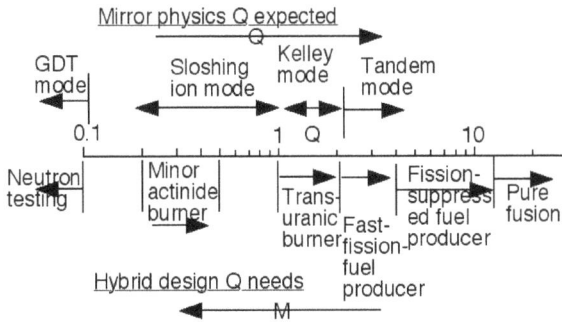

Figure 17.19   Expected mirror physics in tandem mirror [13]

Although mirror fusion physics research has diminished in past decades, recent Russian advances in mirror fusion physics is stimulating renewed interest in the geometrically simpler mirror confinement concept, which is being reconsidered as a FFH concept [13]. The axisymmetric mirror hybrid can accommodate fission blankets designed for any other confinement system. A schematic of a mirror FFH system is presented in Figure 17.16 and its parameters are summarized in Table 17.12.

Table 17.12   Characteristic parameters of a mirror fusion driver for an FFH [10]

Plasma radius[a], m	0.5
Mirror-to-mirror length, m	40
Length of a reacting plasma[2], m	35
Volume of a reacting plasma[2], m^3	25
Plasma surface area[b], m^2	100
Injected ion energy[c], keV	80
Average ion energy[c], keV	40
Average ion density, m^{-3}	10^{20}
Electron temperature, keV	3
Peak ion density, m^{-3}	1.3×10^{20}
$Z_{\text{eff}}^{\text{d}}$	1.2
Magnetic field, T	2.5
Mirror field, T	15
Volume-averaged beta	0.25
s = plasma radius/average ion gyroradius	30
NBI trapped power, MW	65
Plasma Q	0.7
Fusion power, MW	45
Neutron power, MW	36
Neutron wall load, MW/m^2 @ 0.6 m	0.27
Power to end tanks, MW	75

[a]In the midplane.
[b]Between the turning points of the sloshing ions.
[c]Ignoring ½ and 1/3 energies.
[d]Based on the experience with large-scale mirror facilities and composition of the injected particle beams [14].

Based upon recent Russian advances in mirror fusion physics, the magnetic mirror fusion physics concept has been revived in the United States.

17.6   Mirror designs discussion

The Q values required for several different hybrid blankets designed for different purposes are given in Table 17.13. Actinide waste incineration, or burning by fissioning, can be accomplished with fusion neutrons. Blankets can use solid fuel forms or molten salt fuel form. With solid fuel forms, cooling of after-heat requires active or engineered safety systems. By comparison, with molten salt fuel forms, the fuel can be drained passively during off-normal conditions to passively cooled dump or storage tanks (Figure 17.20).

Table 17.13   *Required Q for various versions of the tandem mirror fusion–fission hybrid for the recirculating power fraction = 0.2, $P_{nuclear}$ = 3,000 MW*

Actinide burner			
Blanket multiplication, M	**Minimum Q required**	P_{fusion}, **MW**	**Comments**
Transuranics, $M = 19$	1	200	Solid fuel, engineered or active safety
Minor actinides, $M = 38$–150	0.1–0.5, 0.2 av.	25–100, 50 av.	"
Transuranics, molten salt, $M = 13$	1.5	280	Passive safety
Fuel producer			
Fission suppressed, $M = 2.1$, ^{233}U	8	1,600	Passive safety
Fast fission, $M = 10$, ^{239}Pu	2	370	Engineered safety
Power producer			
$M = 10$	2	370	Molten salt passive safety solid fuel engineered safety
Pure fusion			
$M = 1.34$	I 11	I 2,300	Passive safety

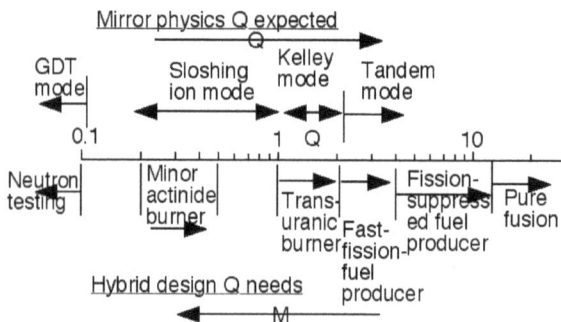

Figure 17.20   *Hybrid options and corresponding mirror operating regimes illustrate the required Q and M tradeoffs [13]*

17.7 Summary of the mirror FFH

A fusion neutron source can be based on an axisymmetric set of magnets employing existing neutral beams at 80 keV to achieve plasma conditions modestly extended from those already achieved. The predicted Q of ~0.7 would be sufficient for applications to burn MA wastes (elements beyond Pu) in the sloshing ion mode. Blanket designs proposed for other fusion concepts could be accommodated even more easily in the axisymmetric mirror owing to its simple geometry. The system would be steady state, requiring neutral beam technology development to extend lifetime to about a year. The heat load is spread over as much area as needed in end tanks and therefore is within the state of the art. Pumping would be by well-known condensation pumping that would need to be made steady by proposed techniques of cycling a portion of the pumps for outgassing at any one time. The solenoidal magnets at 2.5 T are common and even the 15 T mirror magnets are similar to those tested for ITER. With an extra mirror end cell added to each end in the Kelley mode, the Q might be raised to about 1 to permit burning all actinides or in the tandem mode to $Q > 4$ to allow fuel production in the fission-suppressed mode. Fusion applications such as a materials-testing neutron source and other fusion technology will likely be developed independently and can be used by this hybrid application.

Chapter 18

SABR tokamak burner fuel cycles (using fusion neutrons to fission nuclear waste and close the back end of the fission fuel cycle)

Nuclear fission power has an unresolved, but not unresolvable, problem which fusion can at least in large part solve—the disposal of spent nuclear fuel (SNF) containing radioactive transuranics (TRU) with extremely long half-lives of 100,000+ years. While disposal of this "spent" fuel by burial in secured repositories appears to be technically feasible, it is wasteful of an enormous nuclear fuel resource and is adamantly, if irrationally, opposed politically, at least in Nevada in the United States. A more efficient, but more technically difficult, solution of the SNF issue is to separate the long-lived TRU in SNF, the most long-lived of which is mostly fissionable minor actinides (MA), and fission them, thereby not only diminishing the SNF "nuclear waste" problem but also *extracting about 33% more nuclear energy from the nuclear fuel in the process*. What is needed is the neutrons with which to fission these actinides, and these can come from a fusion neutron source, i.e., a fusion–fission hybrid burner reactor.

Such a 3,000-MWth SABR#1 FFH burner reactor has been conceptually designed by the group at Georgia Tech, based on the VTR-IFR fission reactor technology and the ITER fusion technology [44,26]. The geometry is shown in Figures 17.1 and 17.2. The Argonne metal TRU fuel composition of Table 17.4 was used in the design. SABR is fueled with TRU discharged from thermal reactors, initially cast into TRU-Zr metal fuel pins cooled with sodium. The reactor would operate subcritical to achieve a deep-burn four-batch fuel cycle that fissions 25% of the TRU in an 8.2-year residence time, limited by radiation damage accumulated to 200 dpa (displacements per atom) in the oxygen dispersion-strengthened clad and structure. The annual TRU fission rate in SABR (3,000 MWth) is comparable to the annual TRU discharge of three to five 1,000-MWe light water reactors, depending on the plant capacity factor achieved in SABR. A tokamak D–T fusion neutron source, based on the physics and technology that will be demonstrated in ITER, supports the subcritical operation. Several issues related to the integration of fusion and fission physics and technology were identified and subsequently incorporated [26]: refueling a modular, Na-cooled reactor located within the magnetic configuration of a tokamak; achieving long-burn steady-state plasma operation; access for plasma heating and current drive power; suppression of electromagnetic effects in a sodium reactor coolant; refueling the fission reactor; tritium self-sufficiency; shielding of superconducting magnets, etc. [26].

Fuel cycle analyses have been carried out [44, 47]. Fuel residence time in the reactor was assumed to be limited by 200 dpa materials damage to the clad. A 3,000-MWth SABR#1 operating with 75% availability would be capable of burning all of the TRU discharged annually from three 1,000-MWe LWRs, i.e., a SABR#1 would be capable of burning all the MA and unavoidably some of the bred plutonium discharged from three 1,000-MWe LWRs. For the situation where all the plutonium in spent fuel was saved for starting up fast critical reactors, and only the MA processed from "spent" LWR fuel were used to fuel SABRs, a 3,000-MWth SABR#1 could destroy all the MA created annually in 25 LWRs of 1,000 MWe each.

The reduction in the high-level waste repository (HLWR) capacity requirement by this fissioning of the radioactive actinides in spent fuel is calculated to be about a factor of 10, based on the present 100,000-year decay heat limit. This is a major benefit of fusion–fission hybrid reactors.

Thus, one could envision a nuclear fleet with 75% of the energy produced by conventional LWRs (or other types of fission reactors) and 25% of the energy produced by SABRs (or other types of fusion–fission hybrid reactors) that burned all of the TRUs discharged from the LWRS. Alternatively, one could envision a nuclear fleet with 95% of the energy produced by LWRs and 5% produced by SABRs that burned the MA primarily and some small fraction of the plutonium discharged from the LWRs, while plutonium was accumulated to start up a fleet of fast reactors.

Chapter 19

Computational models for tritium breeding ratio and fissile breeding ratios

SABR was modeled in ERANOS (European Reactor ANalysis Optimized calculation System), a fast reactor code system developed to model the Phénix and SuperPhénix reactors. ERANOS employs the European Cell COde (ECCO) to collapse 1968-group JEFF2.0 cross sections within each reactor lattice cell to the 33 groups used in core transport calculations, varying from 20 MeV down to 0.1 eV. The core geometry was described in R-Z cylindrical geometry and the core calculations performed in the ERANOS discrete ordinates transport module BISTRO using an S8 quadrature with 132 radial and 216 axial mesh points. The fuel was depleted for 100 days in each burnup step in the EVOLUTION module before reperforming the core neutron flux calculations.

At each depletion step, the neutron source multiplication k_{mult} was calculated, and the neutron source strength was adjusted such that the fission annulus output would be 3,000 MW(thermal). The fusion power P_{fus} required to maintain a given fission power P_{fis} is determined by k_{mult}, the average number of neutrons released per fission v, the energy released per fusion E_{fus}, and the energy released per fission E_{fis}:

$$P_{fus} = P_{fis} \left(\frac{1 - k_{mult}}{k_{mult}} \right) \times v \left(\frac{E_{fus}}{E_{fis}} \right) \tag{19.1}$$

It is important to note that k_{mult} differs from the more familiar k_{eff}.

The tritium breeding ratio (TBR) is also calculated at each step to determine if enough tritium is being produced to fuel sustained operation of the fusion neutron source. The TBR is defined as

$$\mathrm{TBR}(t) = \frac{\int_V \sum_c^{Li} \phi(r, t) dV}{\int_V S(r, t) dV} \tag{19.2}$$

where S is the fusion neutron source. This only accounts for production of T by ^{6}Li capture and thus is a conservative estimate of the TBR, as T produced in the threshold reaction in ^{7}Li is not counted in the ERANOS calculation. However, since the tritium breeding material is highly enriched in ^{6}Li and the cross section for production via that route is much higher, the approximation should be quite close to the true tritium production rate.

The fissile breeding ratio, FBR, is the instantaneous ratio of the production rate of fissile atoms to their destruction rate, whether through fission or parasitic capture:

$$\text{FBR}(t) = \frac{P(t)}{D(t)} \tag{19.3}$$

The fissile production rate is calculated by integrating the capture rates of the fertile isotopes over the reactor volume:

$$P(t) = \int_V \left(\sum_c {}^{238}\text{U}(r)\phi(r,t) + \sum_c {}^{238}\text{Pu}(r)\phi(r,t) + \sum_c {}^{240}\text{U}(r)\phi(r,t) \right) dV$$

$$\tag{19.4}$$

Though ^{239}Np, rather than ^{238}U, is technically the precursor to ^{239}Pu, ^{239}Np exists in the reactor in a near steady state after its first few half-lives. Thus, by approximately day 20 of fuel residence time, the decay rate of ^{239}Np is equal to the capture rate of ^{238}U. The destruction rate is the volume-integrated absorption rate for all of the fissile isotopes. Only ^{235}U, ^{239}Pu, and ^{241}Pu exist in substantial amounts in the reactor, so other fissile isotopes are omitted from the summation:

$$D(t) = \int_V \left(\sum_c {}^{235}\text{U}(r)\phi(r,t) + \sum_{abs} {}^{239}\text{Pu}(r)\phi(r,t) + \sum_{abs} {}^{241}\text{Pu}(r)\phi(r,t) \right) dV \tag{19.5}$$

Substituting these expressions for the production and destruction rates into Eq. (19.3), we have

$$\text{FBR}(t) = \frac{P(t)}{D(t)}$$

$$= \frac{\int_V \left(\sum_c {}^{238}\text{U}(r)\phi(r,t) + \sum_c {}^{238}\text{Pu}(r)\phi(r,t) + \sum_c {}^{240}\text{U}(r)\phi(r,t) \right) dV}{\int_V \left(\sum_c {}^{235}\text{U}(r)\phi(r,t) + \sum_{abs} {}^{239}\text{Pu}(r)\phi(r,t) + \sum_{abs} {}^{241}\text{Pu}(r)\phi(r,t) \right)}$$

$$\tag{19.6}$$

19.1   Design constraints

There were four hard constraints placed on the reactor design that, if violated, constituted a termination point for that particular trial case. First, the TBR must not fall below 1.15. This "practical" value was chosen from experience because tritium self-sufficiency is a requirement for sustained fusion operation and previous calculations indicate that this excess above unity allows for losses due to inefficiency in tritium collection and for the radioactive decay of any tritium in inventory throughout the operating and refueling cycles. Second, the radiation damage limit of the clad must not be exceeded. The damage limit of ODS MA957 in a fission spectrum is estimated at either 200 DPA or at an accumulated fast fluence of

4×10^{23} n/cm^2. Third, no blanket (fission) zone may surpass 3 at.% burnup, as per the EBR-II Mark-II fuel pin tests. Fourth, no driver fuel may exceed 13.33 at.% burnup; this is reduced from the 20 at.% reached in the IFR pin tests because whereas most fast reactor fuel pins have a plenum-to-fuel volume ratio of unity, the SABrR pins have a ratio of only 2/3.

There were also soft constraints placed on each case, which were considered more as design guidelines. If a soft constraint is violated, the scenario may be continued either if the violation is temporary or if a scenario is approaching the violation of a hard constraint. There were two soft constraints. First, k_{eff} should be significantly below 1 ($k_{eff} < 0.95$ was desired), and the reactor is always very far from prompt critical. Second, the output of the fission core plus blankets should be maintained at 3,000 MW(thermal) using a maximum of 500 MW of fusion power, only 25% more than the ITER design DT fusion power level.

19.2   Results and discussion

19.2.1   TBR case

The configuration of reflector and tritium breeding blanket that emphasizes a high TBR is shown in relation to the fission core and plasma in Figure 19.1. Placing the outboard tritium breeding blanket adjacent to the fission annulus results in a higher neutron capture rate in the blanket than if it were located radially outside the reflector. However, the increase in neutron capture comes at the expense of some of the fissile breeding in the outer radial fissile blanket.

The k_{eff}, k_{mult}, and fusion power required to drive the fission annulus at 3,000 MW(thermal) are shown in Figure 19.2. Shortly before reaching 2,000 days

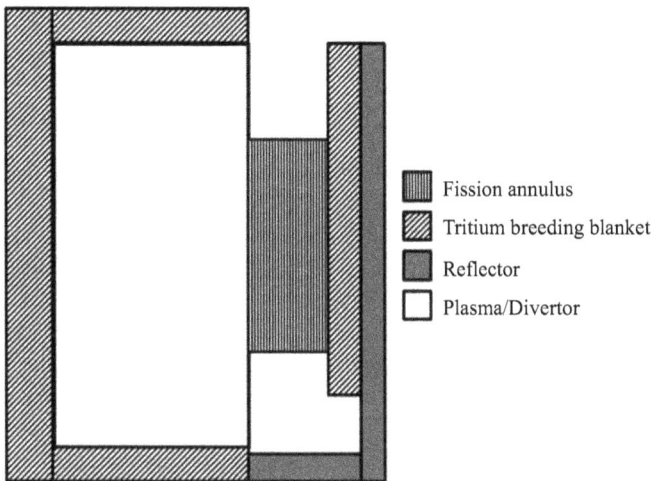

Figure 19.1   Configuration showing fission and tritium breeding structures (other reactor structures omitted)

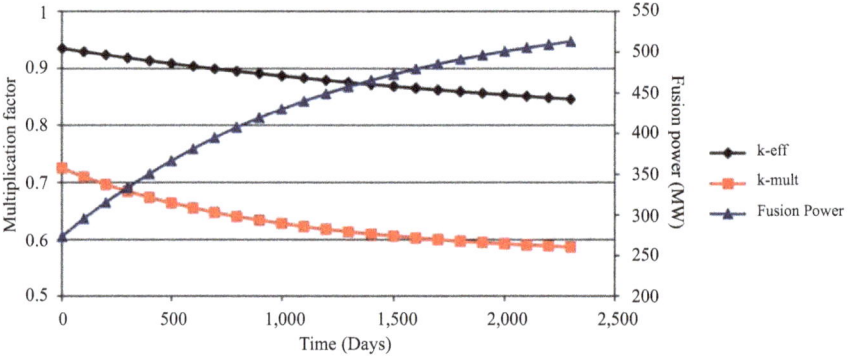

Figure 19.2   Multiplication values and fusion power for TBR case

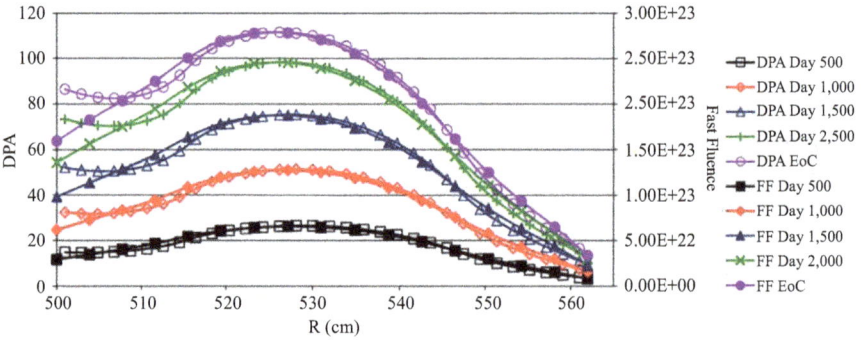

Figure 19.3   Accumulated DPA and fast fluence across core midplane in TBR configuration

of fuel residence time, the fusion power exceeds 500 MW, but the blanket fuel in the plasma-side edge of the inner radial blanket reaches 3% burnup soon after day 2300, at a fusion power of 513 MW. The maximum burnup in the driver fuel is 9.31%, well below its burnup limit of 13.33%. The TBR is substantially above 1.15 for the entire cycle, and the FBR is 1.299 at its peak and 1.278 at the end of the fuel residence time. The average net fissile production over the residence time is 208.4 kg/year.

The fast fluence ($E_n > 0.1$ MeV) and DPA accumulation in the cladding across the fission core midplane are shown in Figure 19.3 at various points throughout the fuel life; the end of cycle (EoC) is after 2,300 days. The contribution of the 14.1-MeV fusion neutrons to the total radiation damage can be seen in the upturn of the DPA curve near $R = 5$ (500 cm). Because these unmoderated fusion neutrons are far more damaging than the average fission neutron in the core, while the fast fluence tallies all neutrons above 0.1 MeV equally, the two curves diverge at the plasma source despite their agreement throughout the rest of the core. This difference is more pronounced later in the core residence time when the source power has been turned up, but the peak for both measures of radiation damage still lies near the midpoint of the fission core and is well below the design limits.

19.2.2   FBR case

A configuration of reflector and tritium breeding blanket that emphasizes a higher FBR can be achieved (Figure 19.4).

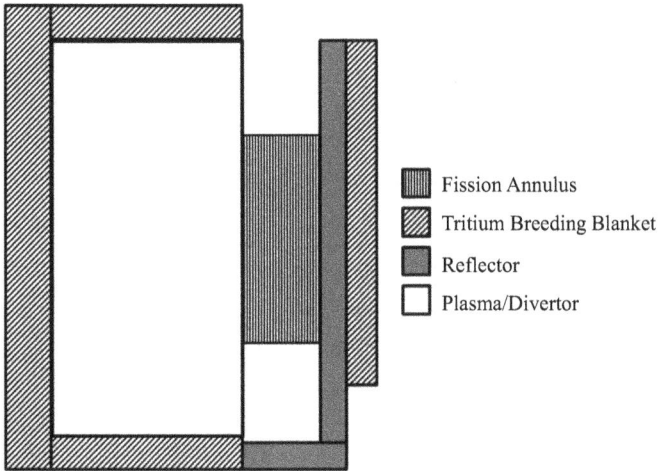

Figure 19.4   Fission breeding structure (FBR) configuration

The k_{eff}, k_{mult}, and fusion power required to drive 3,000 MW(thermal) in the FBR case are shown in Figure 19.5. Higher multiplication values and lower fission power result due to fewer net neutrons leaking radially outward from the outer radial blanket zone than in the case favoring the TBR. This directly causes both more power and more fissile production in that assembly ring and indirectly increases those values in the adjacent driver fuel. The limiting factor for fuel residence time in this configuration is, as in the TBR case, the burnup limit of the plasma-edge blanket fuel being reached. However, because of the relatively lower fusion power throughout the entire residence time and the consequently lower contribution to the 3,000-MW(thermal) fission output from that zone, it took 2,600 days to reach the limit. The driver fuel has a maximum burnup of 10.23%, comfortably below its maximum. The TBR, while lower in this case than in the case favoring tritium production, is 1.206 at its lowest (this occurs at the EoC), with most of the difference resulting from decreased production in the outboard tritium breeding blanket. The FBR is 1.34 at its peak and 1.298 at the EoC. An average net of 253.7 kg/year of fissile material is produced each year in this configuration.

The fast fluence and DPA accumulation across the core midplane are shown in Figure 19.6 at various points throughout the fuel life; the EoC occurs after 2,600 days. Similar to the TBR case, the fusion neutrons cause a divergence of the DPA and fast

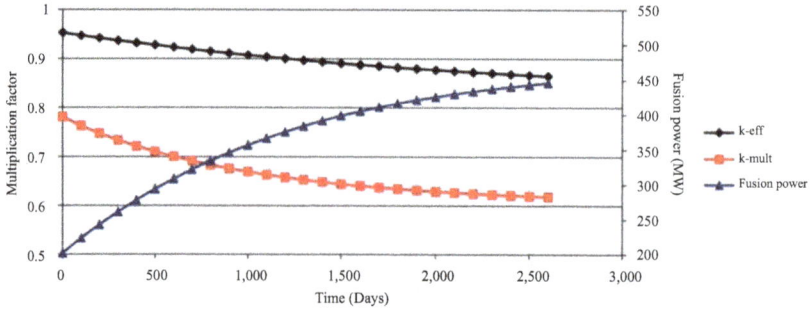

Figure 19.5   Multiplication values and fusion power for FBR case

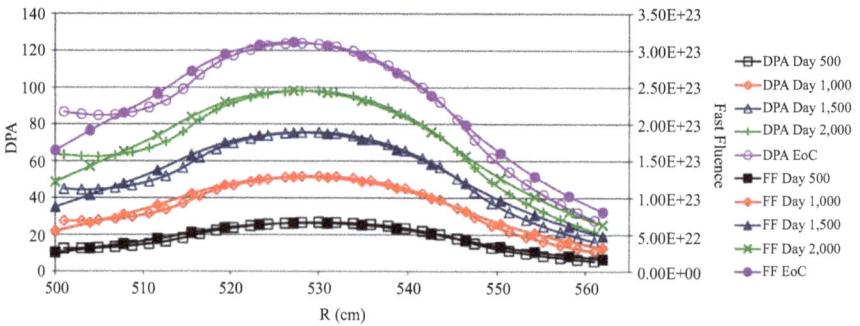

Figure 19.6   Accumulated DPA and fast fluence across core midplane in FBR configuration

fluence near the plasma, but the peaks of both curves are near the annulus center. The maximum DPA is 124, and the maximum fast fluence is 3.12×10^{23} n/cm$^2 \cdot$ s.

19.2.3   Neutronic effect of insulating sheath

A sensitivity study was performed on the FBR configuration to evaluate the effects of removing the LiNbO$_3$ insulating sheath from around each fuel pin. The motivation for doing so stems from the desire to compare with critical fast breeder reactors, which do not require the insulator. While oxide-fueled reactors will have oxygen present in greater fractional quantities than SABrR, Li is absent in even those cores and represents a moderating element unique to SABrR. Though the insulating sheath is less dense than the cladding, it occupies 9.5% of the cross-sectional area within each fuel assembly, so its effect is non-negligible (Figure 19.7).

For this study, the reactor geometry is otherwise identical to the FBR config-uration, and the enrichment of the driver fuel is kept the same. Because the base FBR configuration was able to run until day 2,600 before violating one of the design constraints, the sensitivity study was carried out for the same duration, regardless of violation of design constraints. The effects of removing the sheath from around the fuel pins are summarized in Table 19.1.

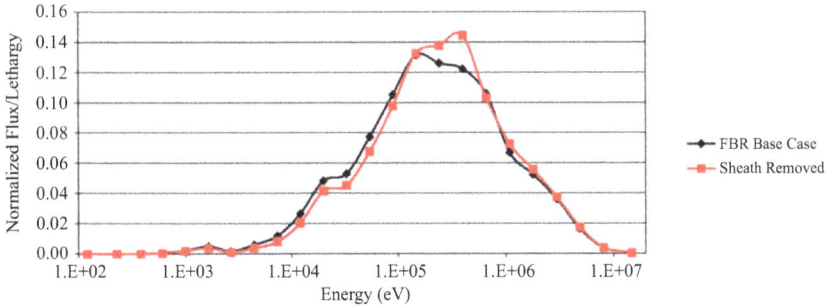

Figure 19.7   Neutron (core-center) spectrum comparison of sheathed and unsheathed pins

Table 19.1   Effects of removing insulating sheath

Quantity	FBR configuration base case	LiNbO$_3$ sheath removed
BoC k_{eff}[a]	0.953	0.971
EoC k_{eff}[b]	0.865	0.879
BoC k_{mult}	0.781	0.870
EoC k_{mult}	0.619	0.666
BoC P_{fus} (MW)	202	108
EoC P_{fus} (MW)	446	364
FBR (peak/EoC)	1.34/1.298	1.301/1.277
Fissile gain (kg/year)	253.7	212.4
TBR (minimum)	1.206	1.366
Peak blanket burnup (at.%)	2.98	2.43
Peak driver burnup (at.%)	10.23	10.48

[a]BoC, beginning of cycle.
[b]EoC, end of cycle.

The removal of the insulating sheath hardened the neutron spectrum. The fission-to-capture ratios of the fuel consequently rose, resulting in a more reactive fission annulus that was easier to drive with the neutron source. The higher fission-to-capture ratios of the fissile isotopes meant a lower destruction rate for a given fission power, which increases the FBR; however, this effect was more than offset by the increase in leakage from the fission annulus, so the resulting FBR is slightly lower due to reduced capture in fertile isotopes. This increase in leakage from the fission annulus is also evident in the higher TBR. Though the demand on the fusion neutron source, and thus the tritium destruction rate, is lower without the sheath, the majority of the tritium production occurs in the blankets surrounding the plasma, whose production is highly dependent on the neutron source strength.

19.2.4   Neutron spectra comparison

The neutron spectra at several points in the core for the TBR and FBR cases are shown in Figures 19.8 and 19.9, respectively, at 1,000 days into the fuel residence

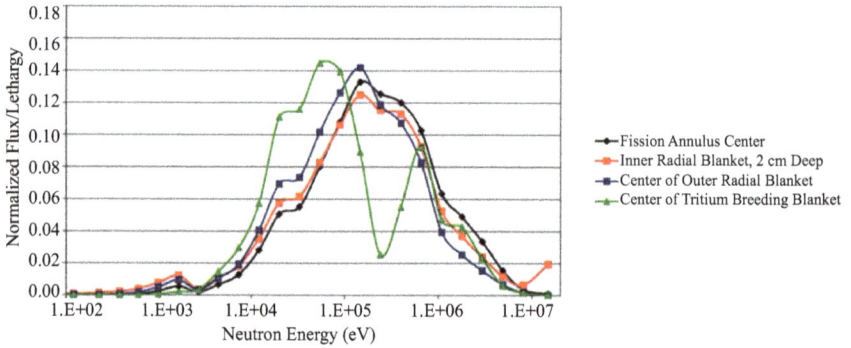

Figure 19.8   Neutron spectra at 1,000 days at selected locations (TBR case)

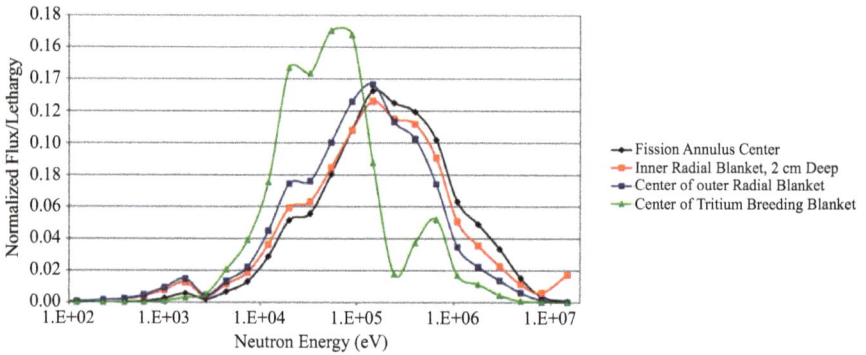

Figure 19.9   Neutron spectra at 1,000 days at selected locations (FBR case)

time; the spectrum in the inner radial blanket only 2 cm away from the plasma demonstrates the effect of the neutron source on the overall spectrum. The spectra are similar throughout the fission annulus; however, in the outboard tritium breeding blanket, there is a pronounced softening of the spectrum. The presence of the reflector between the fission annulus and the tritium breeding blanket in the FBR case significantly enhances this softening of the spectrum relative to the TBR case.

19.2.5   Power distributions

The distribution of power produced in the fusion plasma "driver" and in the fission cores, or "blankets," changes significantly as burnup progresses. Initially, the driver fuel produces nearly all of the fission power, but as fissile isotopes are depleted from the driver fuel and bred in the blankets, the blankets produce an increasing fraction of the power. This increase in blanket power is more pronounced in the inner radial blanket assemblies than the outer ones, as they are exposed directly to the fusion neutron source and thus have a higher rate of breeding and a high incident

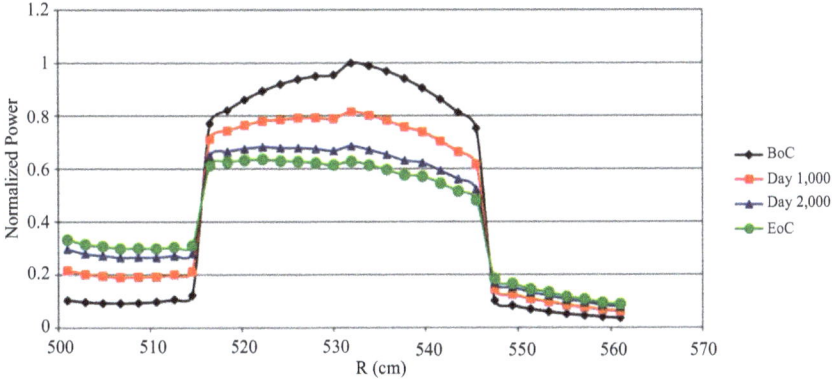

Figure 19.10   Radial power distribution across centerline of the fission annulus at various residence times (FBR case)

neutron flux from the plasma. The radial power distribution for the FBR case is shown for various times in the burnup cycle in Figure 19.10.

The TBR case power distribution is almost identical to that of the FBR case. The power at any given time in the burnup cycle is slightly higher in the inner radial blanket than in the FBR due to the comparatively higher incident fusion neutron flux. The outer radial blanket power is slightly lower in the TBR case due to competing neutron capture in the adjacent tritium breeding blanket suppressing fissile production and reducing the neutron flux in that region.

19.2.6   Comparison of SABrR to critical fast reactor system

A comparison of the breeding performance of SABrR with the high-breeding metal-fueled S-PRISM core design [17] from which the SABrR fuel pins were adapted is shown in Table 19.2. This critical system was chosen for the comparison because of the pin similarity and because of the maturity of the S-PRISM design.

The lower specific power and higher TRU loading of SABrR are a direct consequence of the annular geometry of its fission core; such geometry has a much higher leakage than the traditional pancaked cylinder, so the driver fuel k_∞ must be correspondingly higher, even for a lower k_{eff}. The higher fissile loading of SABrR means that, despite its higher FBR, it has a longer doubling time. However, the fissile gain normalized to fission core thermal power is roughly equal for the SABrR TBR case and S-PRISM, while the SABrR FBR case exceeds S-PRISM in this regard.

SABrR has a higher fuel residence duration than the driver fuel of S-PRISM but a slightly lower blanket residence time. This S-PRISM core design is radially heterogeneous and utilizes blanket shuffling to flatten the radial power profile. SABrR, however, does not shuffle assemblies at any point during the burnup cycle; the presence of the neutron source at the edge of the fission annulus and the ability to adjust its strength largely negate the need to do so for power flattening purposes.

Table 19.2   Breeding performance comparison of SABrR to S-PRISM

Item	SABrR			S-PRISM
Core thermal power (MW)	3,000			1,000
BoC Pu loading (kg)[a]	14,317.0			3,159.9
BoC fissile Pu loading (kg)	9,738.2			2,458.8
BoC U loading (kg)	1,64,763.1			33,052.7
Specific power (W/g Pu)	209.54			316.47
Pu enrich (wt%, Pu/(U + Pu))	Driver zone 1: 22.36			21.29
	Driver zone 2: 23.75			
	TBR case	**FBR case**	**Unsheathed**	—
Fuel residence time (days)	2,300	2,600	2,600	Driver: 2,070
				Blanket: 2,760
Cycle-average breeding ratio	1.28	1.32	1.28	1.22
Fissile gain (kg/year)	208.39	253.73	212.37	69.91
Normalized fissile gain (kg/(MW thermal/year))	0.0695	0.0846	0.0708	0.0699

[a]BoC, beginning of cycle.

*Therefore, SABrR would be shut down far less frequently for shuffling/refueling
purposes, which is an advantage it holds over nearly all critical systems.*
  Because cycle length for both the FBR and TBR cases for SABrR was limited
by blanket burnup in ring 1 with a reasonable margin to peak radiation damage
and driver burnup limits, the radial blankets in rings 1 and 4 might be switched
midcycle to increase total residence time, although at the cost of increased
downtime. Finding a suitable electrically insulating material that has less moderating power than the $LiNbO_3$ would also extend the cycle duration of SABrR
since the blanket burnup limit was not reached in that scenario. A less moderating
insulator would allow for either decreased fissile enrichment of the driver fuel or
for radially heterogeneous core layouts, which do not increase driver enrichment
to high levels.
  We note that no attempt has been made to optimize these initial SABrR
designs for fissile production within a particular fuel cycle. *In principle, subcritical operation (a) removes the criticality requirement, which allows the fuel
and blanket to remain in the reactor until the clad radiation damage limit is
reached, and (b) increases the reactivity margin to prompt critical by an order of
magnitude, which removes any safety limitation on Pu content in the reactor.* A
future investigation will seek to leverage these two factors to improve the fissile
production performance of SABrR for comparison against critical fast burner
reactors. Our purpose here was to investigate whether the SABR burner reactor
concept (technology, geometry, and major parameters) could be adapted to a
breeder reactor that had a reasonable fissile production performance, which seems
to be the case.

19.3 Conclusion

The SABrR FFH fast burner reactor configuration, based on IFR-PRISM fast reactor physics and technology and on ITER fusion physics and technology, was investigated for a fast FFH breeder reactor application. Representative configurations for breeding fissile material from depleted uranium while simultaneously breeding tritium were considered, subject to realistic constraints on (a) the radiation damage to the cladding (200 DPA or 4×10^{23} n/cm^2 fast fluence), (b) driver fuel burnup (13.33 at.%), (c) blanket fuel burnup (3 at.%), and (d) TBR >1.15. *The representative designs considered were found to be capable of producing fissile production/destruction ratios FBR > 1.3, while maintaining TBR > 1.2.* This neutron economy is sufficient to produce 250 kg/year of fissile material in a 3,000-MW (thermal) plant while also producing enough tritium for self-sufficiency of the fusion neutron source fuel.

This study demonstrates the capability of the SABrR FFH fast breeder concept to breed significant excess fissile material and to maintain the tritium needs of the breeding FFH plant. We discuss elsewhere in the book the significant safety advantage of a FFH breeder relative to a similar critical breeder that is associated with the fact that the margin of safety against prompt supercritical runaway is the subcriticality of the FFH reactor, whereas the margin of safety for a critical reactor is the orders of magnitude smaller delayed neutron fraction.

Chapter 20

Physics and engineering design constraints on fusion–fission hybrids

Some insight into the physics and engineering constraints on FFHs can be obtained from relatively simple physics models of global neutron and plasma power balance computational models and models of technological constraints [19–21,26,28–30,32,33] that serve to illustrate how (tokamak) physics and technology limits/constrains the fusion neutron source strength and the level of auxiliary power needed for the neutron sources. We illustrate this by calculating the total neutron population in a nuclear reactor with an external fusion neutron source S using the so-called "point kinetics" model describing the dynamics of the total neutron population "n" in a nuclear reactor with an external source S. Extensions to more sophisticated models are readily envisioned.

We have in mind, of course, that the external neutrons will be from a fusion neutron source. However, these equations do not depend on the type of neutron source, and since an accelerator-spallation neutron source is available and being investigated for this application, we will consider such sources, as well. The following equations do not depend on the type of neutron source, although the design of the reactor of course does. Letting n be the neutron density, C_i be the delayed neutron precursor density of type i, λ_i be the decay rate of precursor C_i into neutrons, S be the external neutron source density, β_i be the delayed neutron fraction of type i, script ℓ be the prompt neutron lifetime in the reactor, and k be the multiplication constant of the reactor, the neutron density in the reactor is governed by the so-called point kinetics equations for the neutron concentration, n

$$\frac{dn}{dt} = \frac{(k-1-\beta)}{\ell}n + \sum_{i=1}^{I} \lambda_i C_i + S \tag{20.1}$$

and the delayed neutron precursor densities are governed by

$$\frac{dC_i}{dt} = \frac{\beta_i n}{\ell} - \lambda_i C_i, i = 1, \ldots, I \tag{20.2}$$

where the C_i represent the population of fission products of species i which decay with half-life $t_{1/2} = \sqrt{2}/\lambda_i$ to release "delayed" neutrons, β_i is the fission yield of delayed neutron precursor species i, $\beta = \sum \beta_i$, k is the multiplication constant of the reactor assembly, and $\ell = 1/(v\Sigma_a)$ is the prompt neutron lifetime, where v is the average neutron speed and Σ_a is the macroscopic absorption cross-section averaged over the neutron energy distribution. The asymptotic solution of these equations yields the equilibrium neutron population density in the reactor,

$$n = \frac{S\ell}{1-k} = \frac{S/v\Sigma_a}{1-k} \tag{20.3}$$

The effective multiplication constant can be written as the ratio of the neutron production rate to the neutron destruction rate,

$$k = \frac{v\Sigma_f + 2\Sigma_{n,2n} + 3\Sigma_{n,3n}}{\Sigma_a^F + \Sigma_a^{Li} + \Sigma_a^P} P_{NL} \equiv \frac{v\gamma\Sigma_f P_{NL}}{\Sigma_a^F + \Sigma_a^{Li} + \Sigma_a^P} \tag{20.4}$$

where v is the average number of neutrons per fission, the factor $\gamma > 1$ takes into account the neutrons produced in $(n,2n)$ and $(n,3n)$ reactions, P_{NL} is the non-leakage probability that a neutron remains in the reactor to be absorbed, and Σ_a^x is the macroscopic absorption cross-section for the fissionable material $(x = F)$, plus the lithium needed to breed tritium for the fusion neutron source $(x = Li)$, plus for absorption in the structure and other parasitic material $(x = p)$.

We assume that in the fusion neutron source it is necessary to produce tritium fuel, which requires a tritium breeding ratio (TBR) greater than unity,

$$TBR = \frac{\Sigma_a^{Li} nv}{S} > 1 \tag{20.5}$$

Combining the above equations yields

$$\Sigma_a^{Li} = \Sigma_a(1-k)TBR \tag{20.6}$$

which can be used with Eq. (20.4) to relate the effective multiplication constant with no Li 9 present (k_0 for TBR = 0) to the effective multiplication constant when Li is present in a reactor that is otherwise identical,

$$k_0 = k[1 + (1-k)TBR] \tag{20.7}$$

The effective transmutation (fission) rate TR is defined as

$$TR = \Sigma_f nv = \frac{S}{(1-k)} \frac{\Sigma_f}{\Sigma_a} \simeq \frac{Sk}{v(1-k)} \tag{20.8}$$

since the non-fission capture of a neutron by an actinide essentially just converts that actinide into another actinide.

20.1   Input electrical power requirement

Since the accelerator-spallation neutron source is also being investigated for use to enhance the number of neutrons in a nuclear reactor, we will also consider it, as well as a fusion neutron source, in this section [27,28].

The input electrical power required to produce S spallation neutrons per second is

$$P_{in,e}^{atw} = \frac{E_{spall}S_{atw}}{\eta_b^{atw}} + P_{aux}^{atw}/\eta_{aux}^{atw} = \frac{E_{spall}S_{atw}}{\eta_b^{atw}}\left[1 + \left(\frac{P_{aux}^{atw}}{E_{spall}S_{atw}}\right)\left(\frac{\eta_b^{atw}}{\eta_{aux}^{atw}}\right)\right] \quad (20.9)$$

where E_{spall} is the energy on target per spallation neutron produced, η_b^{atw} is the ratio of electrical energy to energy on target (i.e., beamline efficiency), P_{aux}^{atw} is the power required to operate the auxiliary systems for the accelerator spallation neutron source transmutation facility, and η_{aux}^{atw} is the efficiency of delivery of this energy to end use. The superscript *atw* refers to "accelerator transmutation of waste".

The input electrical power required to produce S fusion neutrons per second is

$$P_{in,e}^{fus} = \frac{E_{fus}S_{fus}}{\eta_b^{fus}Q_p} + \frac{P_{aux}^{fus}}{\eta_{aux}^{fus}} = \frac{E_{fus}S_{fus}}{\eta_b^{fus}Q_p}\left[1 + \left(\frac{P_{aux}^{fus}}{E_{fus}S_{fus}}\right)Q_p\left(\frac{\eta_b^{fus}}{\eta_{aux}^{fus}}\right)\right] \quad (20.10)$$

where E_{fus}/Q_p is the injected plasma heating energy per fusion neutron produced, η_b^{fus} is the ratio of electrical energy into the heating system to the energy into the plasma (i.e., the heating efficiency), P_{aux}^{fus} is the power required to operate the auxiliary systems for the fusion neutron source transmutation facility, and η_{aux}^{fus} is the efficiency of delivery of this energy to end use.

We will evaluate the ratio of input electrical power required for fusion and accelerator-spallation neutron source transmutation facilities that produce the same transmutation rate:

$$R_{in} \equiv \frac{P_{in,e}^{fus}}{P_{in,e}^{atw}} = \left(\frac{E_{fus}}{E_{spall}}\right)\left(\frac{\eta_b^{atw}}{\eta_b^{fus}}\right)\left(\frac{\xi_{fus}}{Q_p}\right) \times \left\{\frac{\left[1 + \left(\frac{P_{aux}^{fus}}{E_{fus}S_{fus}}\right)Q_p\left(\frac{\eta_b^{fus}}{\eta_{aux}^{fus}}\right)\right]}{\left[1 + \left(\frac{P_{aux}^{atw}}{E_{spall}S_{atw}}\right)\left(\frac{\eta_b^{atw}}{\eta_{aux}^{atw}}\right)\right]}\right\}$$

$$(20.11)$$

$E_{fus} = 17.6$ MeV/n and we use $E_{spall} = 25$ MeV/n, corresponding to the production of 40 spallation neutrons per 1 GeV ion. We assume the same efficiencies for the plasma heating and accelerator systems, so that $\eta_b^{fus}/\eta_b^{atw} = 1.0$. We assume $\eta_b^{fus}/\eta_{aux}^{fus} = \eta_b^{atw}/\eta_{aux}^{atw} = 0.8$. We assume that 50 MW auxiliary power is needed for a transmutation system that produces 2.5×10^{19} n/s and use $P_{aux}/S = 12.5$ MeV/n for both the fusion and atw transmutation systems.

The requirement to breed tritium in the fusion transmutation facility consumes neutrons that could otherwise be used for transmutation. We consider two different scenarios with regard to the treatment of this tritium breeding requirement. In the first scenario, it is assumed that the transmutation facility used with the fusion neutron source differs from the transmutation facility used with the accelerator spallation neutron source only by the addition of sufficient Li to achieve TBR = 1.05. Thus, if the fusion transmutation facility has a multiplication constant k, the ATW facility has a larger multiplication constant k_0 (given by Eq. (20.7)), and the fusion facility must produce a neutron source that is larger by the factor

$$\xi_{\text{fus}} \equiv \frac{S_{\text{fus}}}{S_{\text{spall}}} = \frac{1-k}{1-k_0} = \frac{1+(1-k)\text{TBR}}{(1-k)\text{TBR}} \tag{20.12}$$

in order for both facilities to have the same transmutation rate. We use $\gamma = 1.05$, $v = 2.9$, and $P_{NL} = 0.95$ to calculate k_0 and ξ_{fus}. In the second scenario, we assume that the transmutation reactors used with the fusion and accelerator neutron sources have the same multiplication constant k, and hence the same neutron source levels to produce the same transmutation rates. This would be accomplished by replacing a parasitic absorber with Li or by increasing the fissionable material in the transmutation reactor for the fusion neutron source relative to the ATW facility.

The ratio of input electrical powers required to produce the same transmutation rates in fusion and accelerator-spallation neutron source transmutation facilities is shown as a function of plasma gain for the second of the above Li scenarios in Figure 20.1 for the above parameters and $k = 0:90$. (The results for $k = 0.85$ and 0.95 differ only slightly.) For a plasma gain Q_p greater than about

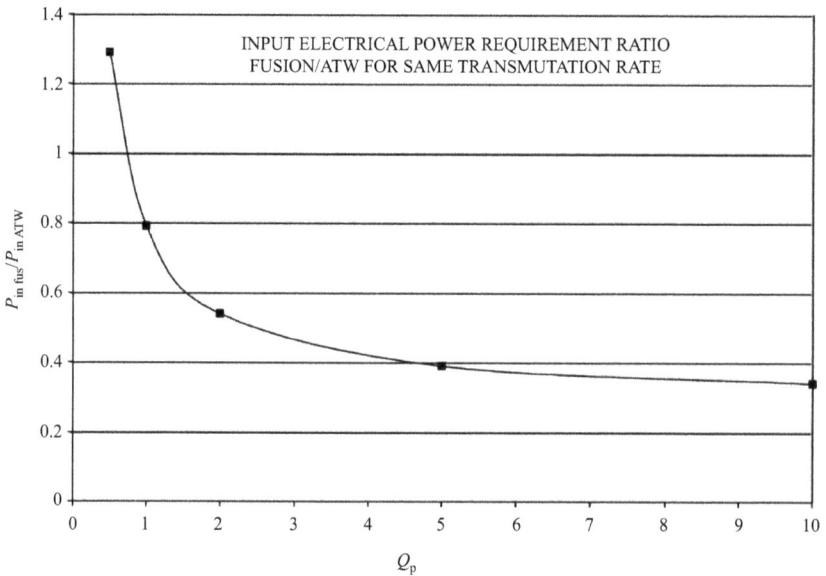

Figure 20.1   Ratio of input electrical power required for the operation of fusion and ATW neutron sources which produce the same transmutation rate in a subcritical transmutation reactor ($E_{\text{spall}} = 25$ MeV/n, $E_{\text{fus}} = 17.6$ MeV/n, $P_{aux}/S = 12.5$ MeV/n, $v = 2.9$, $\gamma = 1.05$, $\eta_b^{\text{fus}} = \eta_b^{\text{atw}} = \eta_e^{\text{fus}} = \eta_e^{\text{atw}} = 0.4$, $\eta_e^{\text{fus}}/\eta_e^{\text{blkt}} = \eta_e^{\text{atw}}/\eta_{aux}^{\text{atw}} = 0.90$, $\eta_b^{\text{atw}}/\eta_{aux}^{\text{atw}} = \eta_b^{\text{fus}}/\eta_{aux}^{\text{fus}} = 0.8 : k = 0.9$ is the same with fusion and ATW neutron sources) [28]

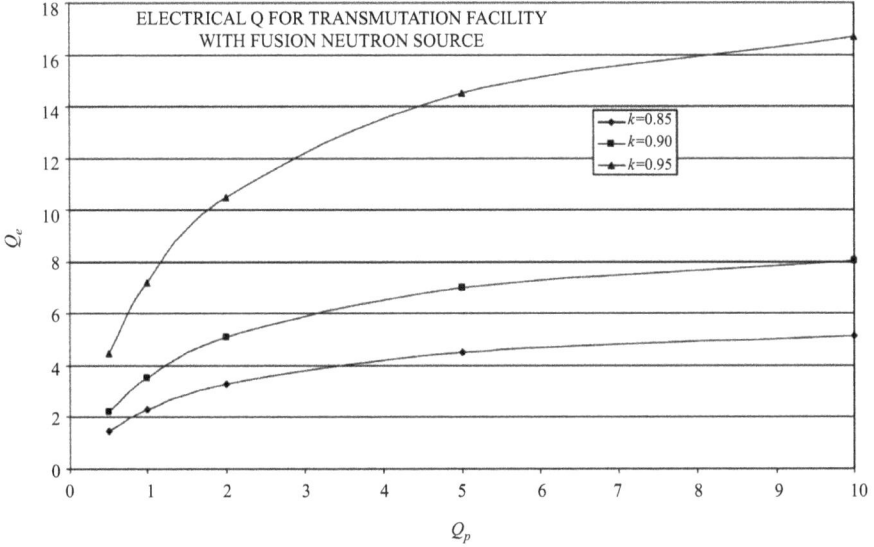

Figure 20.2 *Electrical energy gain Q_e^{fus} for a subcritical transmutation reactor plant with fusion neutron source ($E_{\text{fis}} = 195$ Mev/fis, $E_{\text{fus}} = 17.6$ MeV/n, $P_{\text{aux}}/S = 12.5$ MeV/n, $\nu = 2.9$, $\gamma = 1.05$, $\varepsilon = 0.05$, $\eta_b^{\text{fus}} = \eta_e^{\text{fus}} = 0.40$, $\eta_e^{\text{fus}}/\eta_e^{\text{blkt}} = 0.9$, $\eta_b^{\text{fus}}/\eta_{\text{aux}}^{\text{fus}} = 0.8$) [28]*

1.0–1.5, the required input electrical power is less for a fusion neutron source transmutation facility than for a similar ATW which produces the same transmutation rate. At $Q_p > 5$, the input electrical power requirement for a fusion transmutation facility is about half or less the requirement for an ATW with the same transmutation rate, for the parameters assumed above, and the improvement is slight for $Q_p > 5$.

The electrical energy gain Q_e^{fus} is plotted in Figure 20.2 as a function of the plasma energy gain Q_p. The results shown in Figure 20.2 are in semiquantitative agreement with the results of a similar analysis based on a somewhat simpler model [27].

20.2   Comparison of waste thermal energy

The thermal energy dissipated in the ATW and fusion neutron sources are $E_{\text{spall}}S_{\text{atw}}$ and $(1 + 1/Q_p)E_{\text{fus}}S_{\text{fus}}$, respectively, and the thermal energy generated by the fission and other exoergic processes in the transmutation reactor is $(1 + \varepsilon)E_{\text{fis}}S_x/(1 - k_x)$, where $x = $ fus or atw, and ε is the ratio of energy produced by other (than fission) exoergic reactions to the energy produced by fission. The ratio of the amount of thermal energy that must be removed from transmutation

facilities with fusion and accelerator spallation neutron sources which produce
the same transmutation rate is

$$R_{th} = \left(P_{fus,th}/S_{fus}\right)/\left(P_{atw,th}/S_{atw}\right) = \xi_{fus}\left(1+\frac{1}{Q_p}\right)E_{fus}/E_{spall}$$

$$\times\left\{\frac{\left[1+\frac{(1+\varepsilon)}{(1-k_{fus})}\frac{k_{fus}}{\nu}\left(\frac{E_{fis}}{E_{fus}}\right)\left(\frac{1}{1+\frac{1}{Q_p}}\right)\right]}{\left[1+\frac{k_{atw}}{\nu}\frac{(1+\varepsilon)}{(1-k_{atw})}\left(\frac{E_{fis}}{E_{spall}}\right)\right]}\right\}$$

$$\equiv R_{th}^{source}\frac{\left(\left[1+\frac{(1+\varepsilon)}{(1-k_{fus})}\frac{k_{fus}}{\nu}\left(\frac{E_{fis}}{E_{fus}}\right)\left(\frac{1}{1+\frac{1}{Q_p}}\right)\right]\right)}{\left(\left[1+\frac{k_{atw}}{\nu}\frac{(1+\varepsilon)}{(1-k_{atw})}\left(\frac{E_{fis}}{E_{spall}}\right)\right]\right)} \qquad (20.13)$$

The ratio $R_{th}^{source} = \xi_{fus}\left(1+1/Q_p\right)E_{fus}/E_{spall}$ of the thermal energies dissipated in
the fusion and accelerator neutron sources has been calculated for the two scenarios
discussed earlier for treating the tritium breeding requirement for the fusion neutron
source. A fusion neutron source will dissipate somewhat more thermal energy at low
Q_p (<2) and somewhat less thermal energy at higher Q_p than an accelerator neutron
source, when both sources are driving subcritical reactors with the same k values that
are producing the same transmutation rates ($k_{fus} = k_{atw}$). The thermal energy produced
in the transmutation reactor by fission ($E_{fis} = 195$ MeV/fission) is so much greater than
the thermal energy produced in the spallation neutron source that the ratio of the total
thermal energy produced in fusion and accelerator spallation neutron transmutation
facilities with the same transmutation rate, given by (20.13), is essentially unity.

20.2.1   Comparison of plant electrical energy gain

If the waste energy from the spallation neutron production (i.e., target heating) is
collected and converted to electricity with efficiency η_e^{spall} and the energy of the
fission and other exoergic reactions in the subcritical transmutation reactor is collected and converted to electricity with efficiency η_e^{blkt} the output electrical energy
from a transmutation facility driven by an accelerator spallation neutron source is

$$P_{out,e}^{atw} = \left(E_{spall}\eta_e^{atw} + \frac{k_{atw}}{\nu}\frac{(1+\varepsilon)E_{fis}\eta_e^{blkt}}{1-k_{atw}}\right)S_{atw}$$

where $\varepsilon = 0.05$ accounts for the enhancement of the fission energy release by other
exoergic reactions.

Assuming that the efficiency η_e^{blkt} of collecting and converting to electricity the energy produced in the subcritical reactor assembly is the same with the fusion and accelerator spallation neutron sources, the output electrical energy from a transmutation facility driven by a fusion neutron source is

$$P_{\text{out,e}}^{\text{fus}} = \left(E_{\text{fus}}\eta_e^{\text{fus}}\left(1 + \frac{1}{Q_p}\right) \right) + \frac{k_{\text{fus}}}{\nu}\frac{(1+\varepsilon)E_{\text{fis}}\eta_e^{\text{blkt}}}{1 - k_{\text{fus}}}S_{\text{fis}} \tag{20.14}$$

where η_e^{fus} is the efficiency of collecting and converting to electricity the plasma heating energy and the energy produced by fusion. If the input power, fusion power, and fission power are recovered and converted to electricity, the ratio of the output electrical power to the input electrical power

$$Q_e^x = \frac{P_{\text{out,e}}^x}{P_{\text{in,e}}^x} \tag{20.15}$$

is known as the electrical Q value, or gain, of the system. For a fusion neutron source transmutation facility this expression becomes

$$Q_e^{\text{fus}} = \eta_e^{\text{fus}}\eta_b^{\text{fus}}\left(\frac{\left\{1 + Q_p\left[1 + \frac{k_{\text{fus}}}{\nu}\frac{(1+\varepsilon)}{(1-k_{\text{fus}})}\left(\frac{E_{\text{fis}}}{E_{\text{fus}}}\right)\left(\frac{\eta_e^{\text{blkt}}}{\eta_e^{\text{fus}}}\right)\right]\right\}}{1 + Q_p\left(\frac{P_{\text{aux}}^{\text{fus}}}{S_{\text{fus}}E_{\text{fus}}}\right)\left(\frac{\eta_b^{\text{fus}}}{\eta_{\text{aux}}^{\text{fus}}}\right)} \right) \tag{20.16}$$

We assume the same efficiencies for collecting and converting fusion and spallation heat, so that $\eta_e^{\text{fus}}/\eta_e^{\text{atw}} = 1.0$, and we assume somewhat better heat recovery from the transmutation blanket than from the spallation target or the fusion plasma, so that $\eta_e^{\text{blkt}}/\eta_e^{\text{fus}} = \eta_b^{\text{blkt}}/\eta_e^{\text{atw}} = 0.9$. The quantity Q_e^{fus} is plotted in Figure 20.2, for three different values of the multiplication constant of the transmutation reactor (Figure 20.3), respectively, net electrical power ($Q_e > 1$) is possible for Q_p values even much below unity, for the parameters discussed above.

It can be shown [28] that the ratio of the electrical gains for fusion transmutation and ATW reactor systems which achieve the same transmutation rates is given by

$$\frac{Q_e^{\text{fus}}}{Q_e^{\text{atw}}} = \left(\frac{\eta_e^{\text{fus}}}{\eta_e^{\text{atw}}}\right)\left(\frac{\eta_b^{\text{fus}}}{\eta_b^{\text{atw}}}\right)$$

$$\times \left\{ \frac{\left[\left\{1 + Q_p\left[1 + \frac{k_{\text{fus}}}{\nu}\frac{(1+\varepsilon)}{(1-k_{\text{fus}})}\left(\frac{E_{\text{fis}}}{E_{\text{fus}}}\right)\left(\frac{\eta_e^{\text{blkt}}}{\eta_e^{\text{fus}}}\right)\right]\right\}\right]}{\left[1 + \frac{k_{\text{atw}}}{\nu}\frac{(1+\varepsilon)}{(1-k_{\text{atw}})}\left(\frac{E_{\text{fis}}}{E_{\text{spall}}}\right)\left(\frac{\eta_e^{\text{blkt}}}{\eta_e^{\text{atw}}}\right)\right]} \right\}$$

$$\times \left(\frac{\left[1 + \left(\frac{P_{\text{aux}}^{\text{fus}}}{E_{\text{spall}}S_{\text{atw}}}\right)\left(\frac{\eta_b^{\text{atw}}}{\eta_{\text{aux}}^{\text{atw}}}\right)\right]}{1 + Q_p\left(\frac{P_{\text{aux}}^{\text{fus}}}{S_{\text{fus}}E_{\text{fus}}}\right)\left(\frac{\eta_b^{\text{fus}}}{\eta_{\text{aux}}^{\text{fus}}}\right)} \right) \tag{20.17}$$

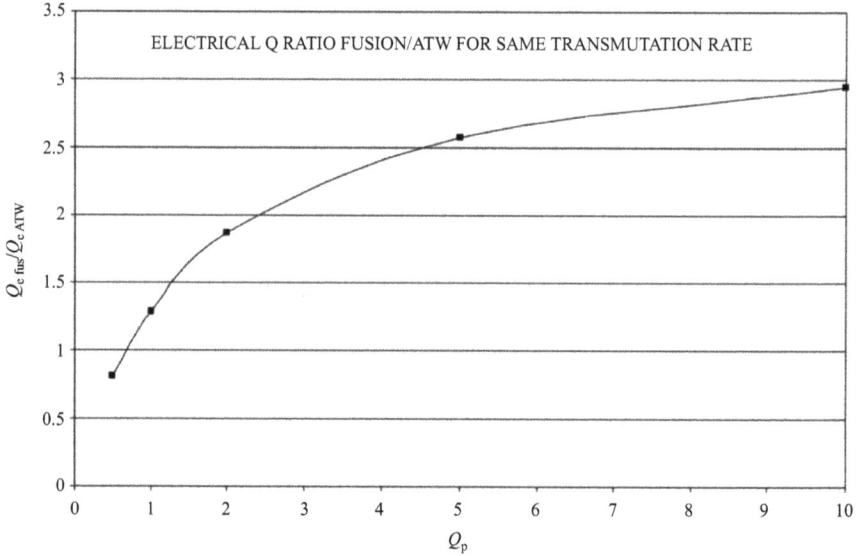

*Figure 20.3   Ratio of subcritical transmutation reactor plant electrical energy
gain for the same transmutation rate with fusion and ATW neutron
sources* $(E_{fis} = 195\text{ MeV}/\text{fis}, E_{spall} = 25\text{ Mev}/\text{n}, E_{fus} = 17.6\text{ MeV}/\text{n},$
$P_{aux}/S = 12.5\text{ MeV}/\text{n}, \nu = 2.9, \gamma = 1.05, \varepsilon = 0.05, \eta_b^{fus} = \eta_b^{atw} =$
$\eta_e^{fus} = \eta_e^{atw} = 0.40, \eta_e^{fus}/\eta_e^{blkt} = \eta_e^{atw}/\eta_e^{blkt} = 0.9, \eta_b^{atw}/\eta_{aux}^{atw} = \eta_b^{fus}/\eta_{aux}^{fus} =$
$0.8, k = 0.90$ *is the same with fusion and ATW neutron sources) [28]*

This ratio is plotted in Figure 20.3 for the parameters described above and for
$k_{fus} = k_{atw}$. For these parameters, the plant electrical gain is greater for fusion
transmutation plants with $Q_p > 1$ than for ATW plants with the same transmutation
rate. These results are in good semiquantitative agreement with the results obtained
in a similar analysis using a somewhat simpler model [27].

We have checked the sensitivity of the calculations to some of the input
parameters. Doubling the auxiliary power per neutron reduces the Q_e ratio by about
33% at high Q_p, but has little effect at low Q_p. Increasing η_b^{atw} from 0.4 to 0.45 or
decreasing η_e^{spall} from 0.4 to 0.35 reduces this ratio by about 15% at high Q_p, but
has little effect at low Q_p, and changing both the ratios $\eta_b^{fus}/\eta_{aux}^{fus}$ and $\eta_b^{atw}/\eta_{aux}^{atw}$ a like
amount has a negligible effect. Thus, the general nature of the comparison of fusion
and accelerator spallation neutron source transmutation facilities shown in
Figure 20.3. These figures would seem to be relatively insensitive to modest
uncertainties in plant parameter values, although a parameter such as the electrical
Q_e for either system will vary linearly with η_e and η_b.

We recognize that a comparison of transmutation efficiency and neutron cost,
as well as a comparison of electrical and thermal power characteristics, is needed to
fully evaluate a fusion neutron source vis-à-vis an accelerator neutron source for
spent fuel transmutation. However, a comparison of efficiency and neutron cost

depends not only on the neutron source but also on the subcritical transmutation facility and is beyond the scope of the present book, but the above discussion indicates what must be involved in such an evaluation.

20.2.2 Limitations on fusion neutron source capabilities for transmutation reactors

The factors determining the effect of possible neutron source limitations on the transmutation rate in a subcritical transmutation reactor can be understood by examining Eq. (20.8). This equation can be put into different forms to illustrate the effect of various possible limitations of the neutron source capability on the transmutation rate (Figure 20.4). We stress that we are examining only limitations on transmutation rate that would be caused by limitations in neutron source capability, not by engineering and safety constraints on the transmutation reactor itself.

20.2.3 Radiation damage limits to the first wall

Writing the neutron source in terms of the power P_n of $E_n = 14$ MeV neutrons through the first wall relates the neutron source strength to the neutron wall load, or

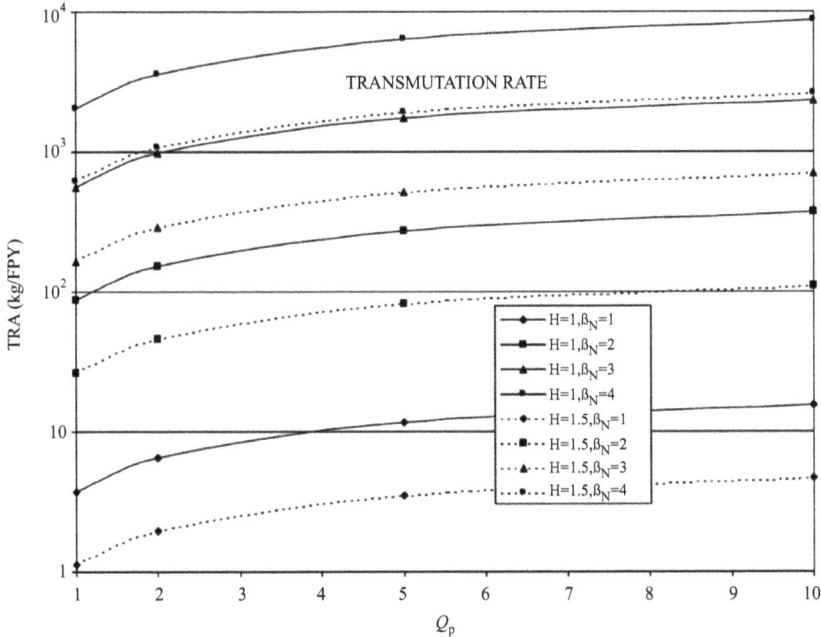

Figure 20.4   Transmutation rate in a subcritical transmutation reactor driven by a tokamak fusion neutron source $(k = 0.90, B = 5.5$ T, $q_{95} = 4, \tau_{ign} = 5$ s, $T = 12$ keV) [28]

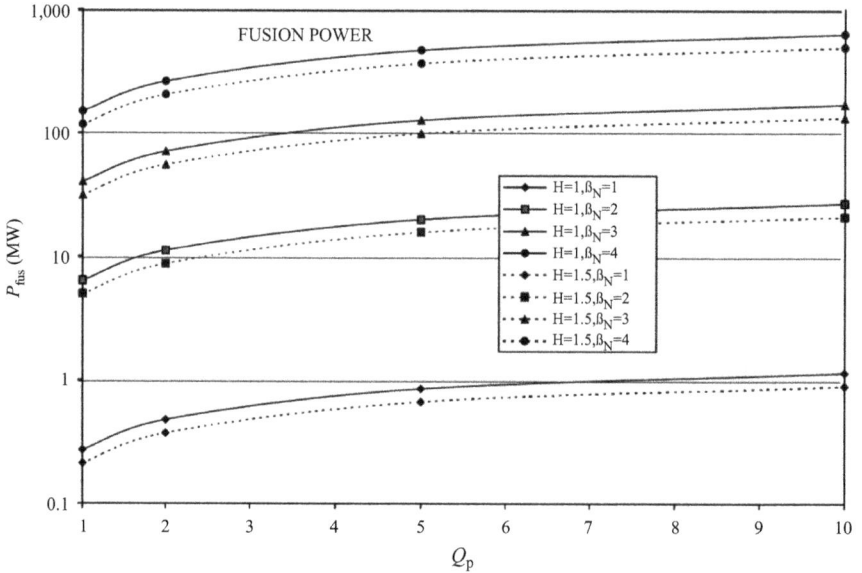

Figure 20.5   Fusion power in tokamak fusion neutron sources for subcritical transmutation reactors $(k = 0.90, B = 5.5 \text{ T}, q_{95} = 4, \tau_{\text{ign}} = 5 \text{ s}, T = 12 \text{ keV})$ *[28]*

flux, Γ_n and first wall area A_w,

$$S = \frac{P_n}{E_n} = \frac{\left(\frac{P_n}{A_w}\right) A_w}{E_n} = \frac{\Gamma_n^{\text{av}} A_w}{E_n} \tag{20.18}$$

This relation may be used to write the transmutation rate as

$$\text{TR} = \frac{S}{1-k}\left(\frac{k}{v}\right) = \frac{\Gamma_n^{\text{av}} A_w \left(\frac{k}{v}\right)}{E_n(1-k)} \tag{20.19}$$

from which the annual transmutation rate TRA (kg/FPY) may be written in terms of the annual fluence, the area of the first wall, and the effective multiplication constant of the transmutation reactor TRA:

$$\text{TRA}\left[\frac{g}{\text{FPY}}\right] = \frac{5.55\frac{k}{v}\left(\Gamma_n^{\text{av}} t\right)\left[MW - \frac{\text{FPY}}{m^2}\right] A_w[m^2]}{1-k} \tag{20.20}$$

where FPY refers to full power year. A given transmutation rate can obviously be achieved in a number of ways by making trade-offs between neutron wall load and first wall area on the one hand and the effective multiplication constant on the other. The upper limit on the effective multiplication constant is set by safety considerations; k should be sufficiently less than unity that no credible event could cause a reactivity increase k that would make the transmutation reactor prompt critical; i.e., $k + \Delta k$ must remain less than $1 + \beta$ for any credible accident, where β is the delayed neutron fraction ($\beta = 0.0020$ for ^{239}Pu and 0.0054 for ^{241}Pu). The wall area depends on the size of the neutron source, of course, which can be made as large as desired for a fusion neutron source.

The first wall will fail when the accumulated neutron fluence $\Gamma_n^{max} t$ reaches some limiting value (1–3 MW FPY/m^2 for existing austenitic stainless steel and considerably more for advanced structural alloys presently under development). A practical requirement might be that the first wall lifetime be as long as the refueling interval for the transmutation reactor. Assuming a 1 FPY refueling interval and taking a limiting neutron fluence of 2 MW FPA/m^2 and a first wall area of 450 m^2 (representative of an $R \approx 5$ m tokamak), a transmutation reactor with $k = 0.90$ driven by a fusion neutron source would be limited by materials damage to a transmutation rate of TRA $\approx 50,000$ kg/FPY. Assuming a first wall area of 100 m^2 (perhaps representative of advanced confinement concepts in the early stages of development), radiation damage in an austenitic stainless steel first wall would limit the transmutation rate to TRA $\approx 10,000$ kg/FPY. Such large transmutation rates are unrealistic for other reasons (handling the engineering challenges of such a large thermal power output and the safety challenges of such a large mass of fissile material in the transmutation reactor being foremost among them). As a point of reference, the present ATW plan is to fission less than 2,000 kg/FPY of actinides in a single transmutation reactor, and studies of critical fission transmutation reactors typically have transmutation rates of less than 1,000 kg/FPY. The main points are that:

(a) A practical fusion neutron source is unlikely to be limited by radiation damage to the first wall, even for the presently available austenitic stainless steels;
(b) A fusion neutron source for transmutation could meet its objectives operating at rather low neutron wall loads ($\ll 1$ MW/m^2).

We will make a more quantitative investigation of these points later in this section.

20.2.4   Thermal limits to the first wall

The thermal power P_{th} which must pass through the first wall of the fusion neutron source is the sum of $E_\alpha = 3.5$ MeV alpha particle energy per fusion and the input plasma heating power, which may be related to the fusion power $P_{fus} = SE_{fus}$ by $Q_p = P_{fus}/P_{heat}$. A fraction f_{div} of this power will be exhausted in the divertor, and a fraction, $(1 - f_{div})$ will be incident on the first wall of the neutron source as a heat flux. Thus, the fusion neutron source may be written as

$$S = \frac{P_{th}}{\left(E_\alpha + \frac{E_{fus}}{Q_p}\right)(1 - f_{div})} = \frac{\Gamma_{th} A_w}{E_{fus}\left(\frac{E_\alpha}{E_{fus}} + \frac{1}{Q_p}\right)(1 - f_{div})} \qquad (20.21)$$

and the corresponding form for the transmutation rate becomes

$$TR = \frac{\left(\frac{k}{\nu}\right)S}{1 - k} = \frac{\left(\frac{k}{\nu}\right)\Gamma_{th} A_w}{E_{fus}\left(\frac{E_\alpha}{E_{fus}} + \frac{1}{Q_p}\right)(1 - f_{div})(1 - k)} \qquad (20.22)$$

In terms of transmutation rate per FPY this becomes

$$TR(kg/FPY) = \frac{4.44\left(\frac{k}{\nu}\right)\Gamma_{th}[MW/m^2]A_w[m^2]}{\left(\frac{1}{Q_p}+\frac{1}{5}\right)(1-k)(1-f_{div})}$$ (20.23)

The maximum surface heat flux for an austenitic stainless steel first wall is about 0.5 MW/m². Taking an average heat flux of half this value to account for peaking, assuming a diverted heat fraction of $f_{div} = 0.5$ and assuming $Q_p = 5$, $A_w = 450\,m^2$ and $k = 0.90$, the transmutation rate would be limited to about 25,000 kg/FPY by first wall heat flux removal limitations. Assuming instead $A_w = 100\,m^2$, the first wall thermal limit on the transmutation rate for austenitic steel is about 5,500 kg/FPY. Again, these are quite large transmutation rates, and the point is that thermal limitations on an austenitic stainless steel first wall of the neutron source should not limit the transmutation rate achievable in a transmutation reactor driven by a fusion neutron source. This point will be investigated more quantitatively in the next section.

20.2.5   Tokamak physics limits

The plasma physics of the fusion neutron source also imposes certain constraints on the realizable source performance. The existing physics basis for a tokamak neutron source is well characterized by the tokamak physics database compiled for ITER. For our purposes, we can consider five such physics constraints to characterize the physics limitations on a tokamak neutron source.

Assurance of MHD stability of existing tokamak plasmas can be characterized by the requirement that the normalized beta parameter β_N

$$\beta_N = \frac{\beta_\%}{I_{MA}/a_m B_T}$$ (20.24)

does not exceed 2.5–3.0 and that the safety factor evaluated at the 95% flux surface

$$q_{95} = \frac{5B_T R_m}{2A^2 I_{MA}}\left[1 + \kappa^2\left(1 + 2\delta^2 - 1.2\delta^3\right)\right] \times \left(\frac{1.17 - 0.65/A}{1 - A^{-2}}\right)$$ (20.25)

be not less than about 3. Here B_T is the toroidal magnetic field in Tesla, I_{MA} is the plasma current in MegaAmps, $A = R/a$ is the ratio of the plasma major to minor radii, κ is the elongation of the plasma, δ is a parameter known as the triangularity which characterizes the degree of D shape of the plasma cross-section, and β is the ratio of the plasma kinetic pressure to the pressure of the magnetic field,

$$\beta = \frac{\left(n_{DT} T_{DT} + n_e T_e + n_{imp} T_{imp} + n_f E_f\right)}{B^2/2\mu_0}$$ (20.26)

Here E_f is the energy of fast plasma ions in the process of thermalizing.

The DT ions, electrons (e), impurities (imp) and fast (f) beam, and alpha particles contribute to the kinetic pressure of the plasma. The plasma must have adequate energy confinement to achieve a power balance between the self-heating from the fusion alpha particles plus any auxiliary (NBI or RF) heating and losses due to transport and radiation. Consideration of the plasma power balance indicates that the energy confinement time must be about $\tau_{ign} = 5$ s to maintain the power balance at about 10 keV average temperature in the absence of auxiliary heating and that the energy confinement time must be about $\tau_{ign}/(1 + 5/Q_p)$ when auxiliary heating is present (Figures 20.6 and 20.7). The tokamak database for ELMy H mode discharges is well correlated to the tokamak parameters by the ITER-98H scaling law $\tau = \tau_{98}(I, B, P_{fus}, \ldots)$. Thus, achieving sufficient energy confinement to maintain power balance imposes the confinement constraint

$$\frac{\tau_{ign}}{1 + \dfrac{5}{Q_p}} = 0.05\, H I_{MA}^{0.91} B_T^{0.15} n_{19}^{0.44} \left[P_{fus} \left(\frac{1}{5} + \frac{1}{Q_p} \right) \right]^{-0.65} \times R^{2.05} \kappa^{0.72} M^{0.13} A^{-0.57}$$

(20.27)

where H is the confinement enhancement factor, M is the plasma ion mass in amu, A is the ratio of the major to minor axes of the tokamak, n^{19} is the density in units of

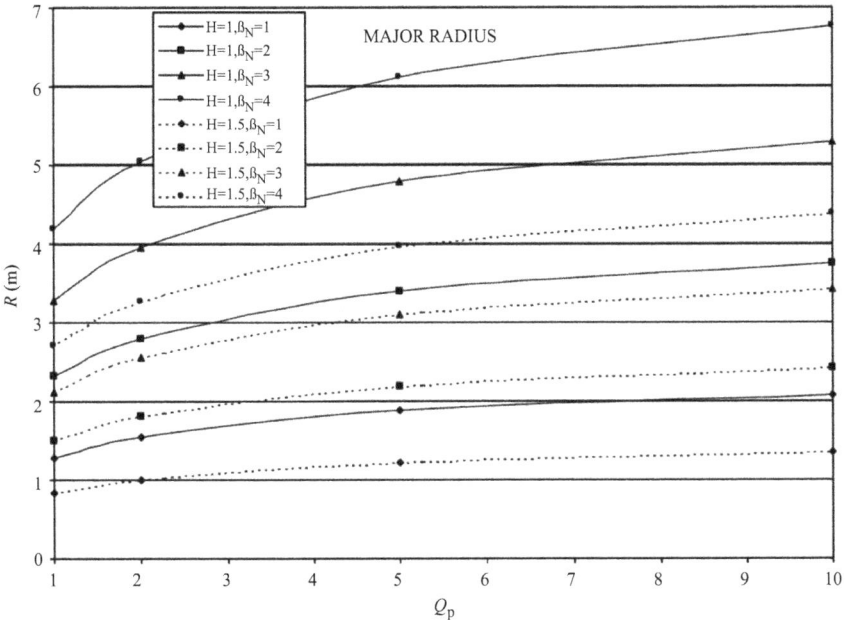

Figure 20.6 Major radii of tokamak fusion neutron sources for subcritical transmutation reactors $\left(k = 0.90, B = 5.5 \text{ T}, q_{95} = 4, \tau_{ign} = 5 \text{ s}, T = 12 \text{ keV} \right)$ [28]

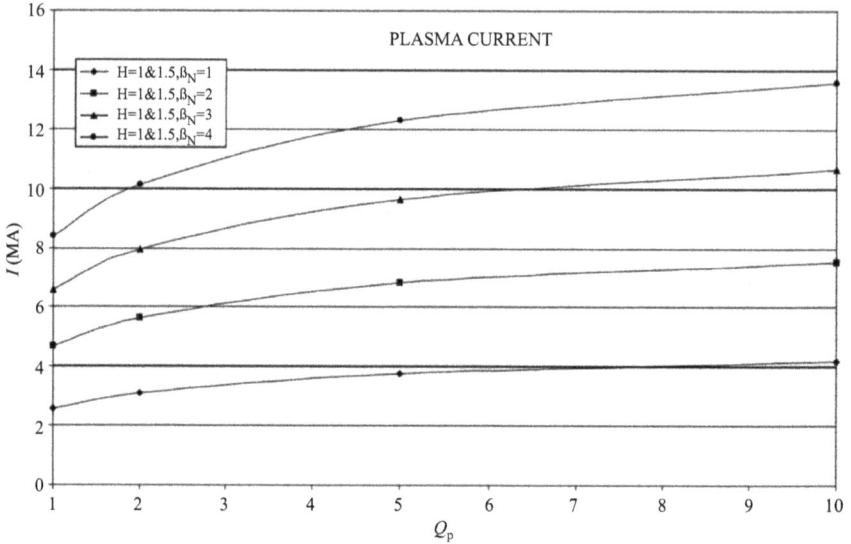

Figure 20.7   Plasma current in tokamak fusion neutron sources for subcritical transmutation reactors $(k = 0.90, B = 5.5$ T, $q_{95} = 4, \tau_{ign} = 5$ s, $T = 12$ keV) [28]

10^{19} m³, P_{fus} is in megawatts and the units of the other quantities are indicated by the subscripts.

The fusion power

$$P_{fus} = \frac{1}{4}n_{DT}^2 \sigma v U_\alpha (\pi\kappa a^2)(2\pi R) \tag{20.28}$$

not only depends on the density, which is constrained by the MHD and confinement constraints, but is also involved in the confinement constraint, so that Eq. (20.28) is actually a fourth physics constraint. Here $\langle\sigma v\rangle$ is the Maxwellian averaged fusion rate and $U_\alpha = 3.5$ MeV is the energy of the DT fusion alpha particle which remains in the plasma to heat it.

A fifth physics constraint, thought to be due to thermal instabilities in the plasma, is an upper limit on the achievable plasma density. This upper limit on the density can be correlated remarkably well over a wide range of tokamak experiments with the simple "Greenwald limit,"

$$n \le n_{GW} \equiv \frac{I_{MA}}{\pi a_m^2} \tag{20.29}$$

although there are many examples of tokamak operation well above (up to about twice) the Greenwald limit (Figure 20.8).

In order to understand how these physics limits might constrain the performance of a fusion neutron source for a subcritical transmutation reactor, we have fixed various parameters $(B = 5.5, q_{95} = 4, A = 3, T = 12$ keV) and solved

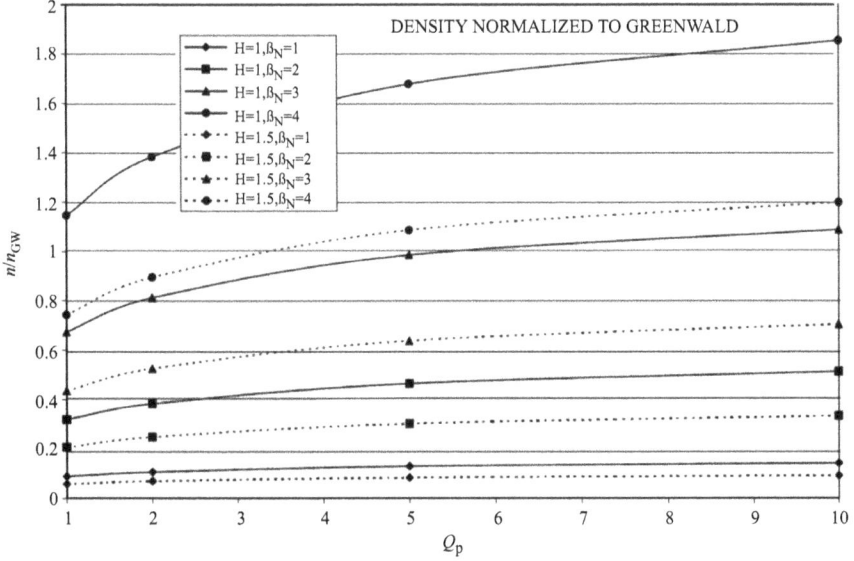

Figure 20.8 *Greenwald-normalized densities in tokamak fusion neutron sources for subcritical transmutation reactors ($k = 0.90, B = 5.5$ T, $q_{95} = 4$, $\tau_{\text{ign}} = 5$ s, $T = 12$ keV) [28]*

Eqs. (20.24, 20.25, 20.27 and 20.28) for $(I, n, R, P_{\text{fus}})$ as a function of Q_p and δ_N. For this purpose we assumed $\beta = 2.1\, n_{\text{DT}} T_{\text{DT}}/(B^2/2\mu_0)$. We carried out the calculations with a set of confinement and shape parameters characteristic of the present tokamak database ($H = 1, \kappa = 1.7, \delta = 0.5$) and with a second "advanced" set characteristic of improved confinement regimes presently under intensive investigation and a higher degree of shaping ($H = 1.5, \kappa = 2.0, \delta = 0.8$). We considered several values of the parameter β_N, including values within the presently established database of $\beta_N < 3$ and with values with the advanced range $\beta_N > 3$ that is presently under investigation. The results are shown in Figure 20.8.

In order to relate these calculations of physics performance to the transmutation rate, we write

$$\text{TRA} = \frac{(k/\nu)S}{1-k} = \frac{4.44(k/\nu)P_{\text{fus}}(MW)}{1-k}\,(kg/\text{FPY}) \qquad (20.30)$$

We choose $k = 0.90$ for the calculations because this value of the multiplication constant provides a large margin of safety to criticality and allows for the processing of spent fuel with large concentrations of parasitic absorbers present. The transmutation (fission) rate is, of course, directly related to the fission thermal power production rate in the transmutation reactor:

$$P_{\text{fis}} = \frac{k}{\nu}\frac{P_{\text{fus}}}{(1-k)}\frac{E_{\text{fis}}}{E_{\text{fus}}} = \frac{11.1}{1-k}\frac{k}{\nu}P_{\text{fus}} \qquad (20.31)$$

Each kg/FPY transmutation rate corresponds to a fission thermal power production rate of about 2.48 MW, when $k = 0.90$.

In order to relate the physics performance to the neutron wall load/radiation damage limit and the thermal limit on the first wall discussed above, we write

$$\Gamma_n = \frac{\frac{4}{5} P_{fus}}{\sqrt{\frac{1}{2}(1 + \kappa^2)}(2\pi a)(2\pi R)} \tag{20.32}$$

$$\Gamma_{th} = \frac{\frac{1}{5} P_{fus}(1 - f_{div})}{\sqrt{\frac{1}{2}(1 + \kappa^2)}(2\pi a)(2\pi R)} \tag{20.33}$$

and carry out the calculation for $f_{div} = 0.5$.

The results of these calculations, which are plotted in Figures 20.9 and 20.10 lead to several interesting conclusions. Foremost among these is the conclusion that parameters routinely achieved in operating tokamaks (i.e., those that are part of the present tokamak database ($H = 1, \beta_N = 2-3$)) are sufficient for a tokamak neutron source operating with densities below the Greenwald limit to produce transmutation rates of several hundred to several thousand kg/FPY in a reactor with $k = 0.9$. Such tokamak neutron sources would have major radii in the range $R = 3-5$ m,

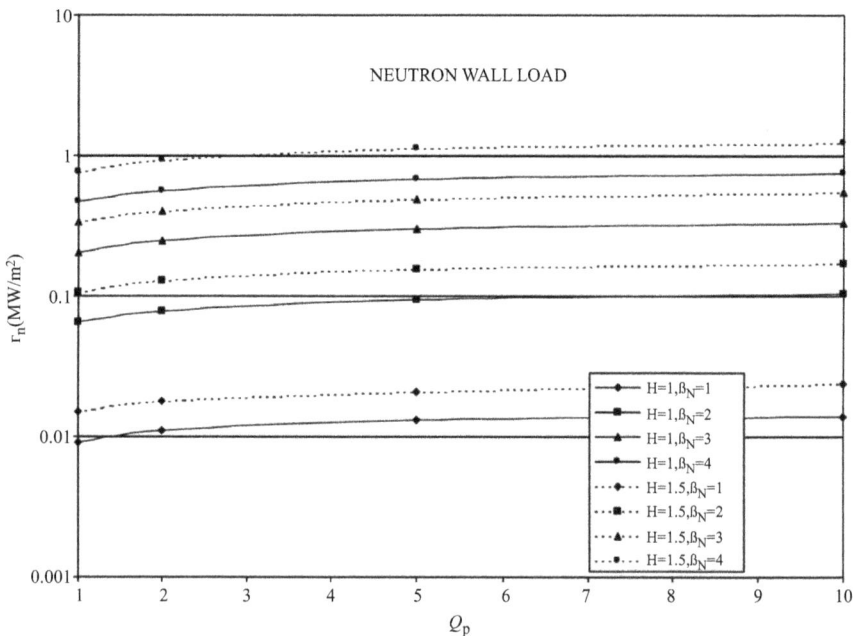

Figure 20.9   Neutron wall loads in tokamak fusion neutron sources for subcritical transmutation reactors $\left(k = 0.90, B = 5.5\ \text{T},\ q_{95} = 4, \tau_{ign} = 5\ \text{s},\right.$ $T = 12\ \text{keV}\left.\right)$ [28]

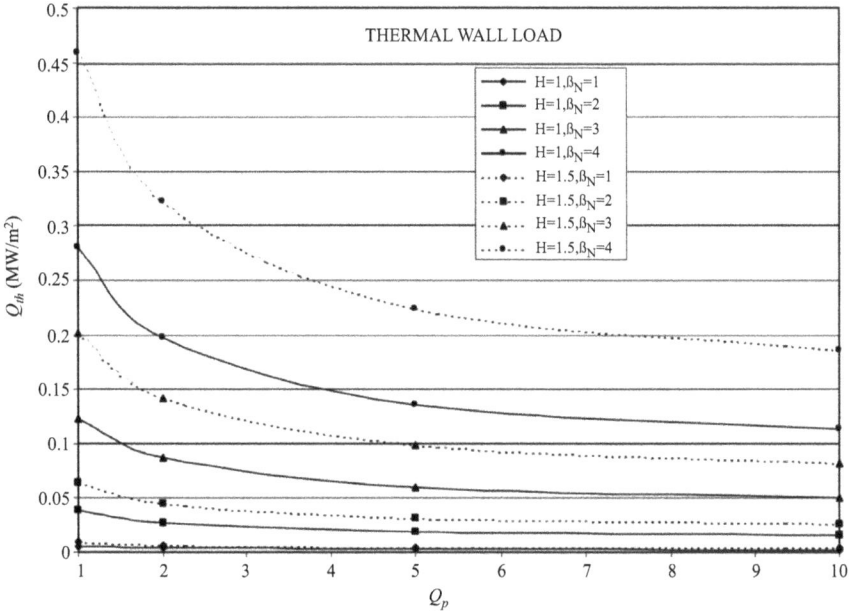

Figure 20.10   Thermal wall loads in tokamak fusion neutron sources for subcritical transmutation reactors ($k = 0.90$, $B = 5.5$ T, $q_{95} = 4$, $\tau_{ign} = 5$ s, $T = 12$ keV) [28]

produce fusion power in the range $P_{fus} = 10{-}100$ MW, and drive transmutation reactors that produce fission thermal power in the range $P_{fis} \approx 1{,}000{-}10{,}000$ MW. Achieving improved confinement $(H = 1.5)$ will enable these same transmutation rates to be achieved with smaller (and presumably less expensive) neutron sources.

As a point of reference, the proposed ATW plant would use two 45 MW proton beams to produce 840 MW thermal energy in each of eight target transmutation assemblies, for a total thermal power output of 6,720 MW, in order to produce a transmutation rate of 1,760 kg/FPY. A second important conclusion is that $Q_p = 2{-}5$ is adequate for the neutron source, since the transmutation rate increases only slowly above $Q_p = 5$ but drops sharply below $Q_p = 2$. Furthermore, the input electrical power requirement and the waste heat production rate for a fusion neutron source, relative to an accelerator spallation neutron source, both become much more favourable above about $Q_p = 1.5{-}2$ but are relatively insensitive to further increases in Q_p beyond about 5.

A third important conclusion is that $\beta_N = 2{-}3$ seems to be the correct range for achieving these interesting transmutation rates (with $k = 0.9$), since $\beta_N = 1$ leads to transmutation rates that are too low to be interesting and $\beta_N = 4$ leads to transmutation rates that are so large as to imply large size, fissile inventory, and heat removal challenges in the design of the transmutation reactor. We note in this regard that since the transmutation rate shown in Figure 20.4 scales as $(1 - 0.9)k/0.9(1 - k)$, interesting transmutation rates could also be achieved at $\beta_N < 2$ in transmutation reactors with $k > 0.9$ and at $\beta_N > 3$ in transmutation reactors with $k < 0.9$.

20.2.6   Engineering limits on a tokamak fusion neutron source

The smaller end of the size range indicated would have to use copper magnetic technology rather than superconducting magnet technology, and even then some of the cases shown may be excluded by engineering limits. For example, at $R = 2$ m and minor radius $a = 0.67$ m $(A = 3)$, a 5-MA inductive current capability (for backup and testing) would require about 8.5 V/s, which would necessitate a flux core radius $a_{fc} \approx 0.5$ m at a maximum field of 10 T in the ohmic heating coil, leaving only $2 - 0.67 - 0.5 = 0.83$ m for the ohmic heating coil, the toroidal field coil, and any inboard section of the transmutation reactor, which at most would be just a very thin neutron shield. (See [28] for a discussion of the engineering limitations on the size of tokamaks.) With such a small-R neutron source, the fraction (perhaps 20%) of the neutrons directed inwards (toward the major axis of the tokamak) would be absorbed in the copper magnets and shield, effectively reducing the neutron source to the transmutation reactor proportionately.

At $R = 4$ m and $a = 1.33$ m, a 5-MA inductive current capability would require about 35 V/s, which would necessitate $a_{fc} \approx 1.05$ m, leaving about 1.6 m on the inboard side for the ohmic and toroidal coils and the shielding. This is still neither enough space on the inboard side for shielding to enable use of superconducting magnet technology nor enough space for placing transmutation assemblies on the inboard $\approx 20\%$ of the neutron source. Thus, the effective neutron source for transmutation would again be reduced by $\approx 20\%$.

Table 20.1   *Characteristic parameters of tokamak fusion neutron sources that produce transmutation rates of hundreds to thousands of kg/FPY in transmutation reactors with $k = 0.9$*

Parameter	Nominal database	Advanced database
H confinement	1.0	1.5
β_N	2–3	2–3
Q_p	2–5	2–5
$B(T)$	5.5	5.5
$\tau_{ign}(s)$	5	5
κ, δ	1.7, 0.5	2.0, 0.8
q_{95}	4	4
n/n_{GW}	0.4–1.0	0.2–0.6
$R(m)$	3–5	2–3
$I(MA)$	6–10	6–10
$P_{fus}(MW)$	10–100	10–100
$\Gamma_n(MW/m^2)$	0.1–0.3	0.15–0.5
$\Gamma_{th}(MW/m^2)$	0.02–0.09	0.03–0.15

At $R = 5$ m and $a = 1.67$ m, a 5-MA inductive current capability would require about 44 V/s, which would necessitate $a_{fc} \approx 1.18$ m, leaving about 2.15 m on the inboard side for ohmic and toroidal coils and an inboard transmutation blanket or shield. At this size, superconducting technology could be used if a shield was placed on the inboard $\approx 20\%$ of the tokamak surface, once again reducing the neutron source to the transmutation reactor by $\approx 20\%$. Alternatively, copper magnet technology could be used and some of this inboard volume could be occupied by transmutation assemblies.

Thus, the effective neutron source to the transmutation reactor would be reduced by as much as about 20% by engineering limitations on the inboard "radial build" of the tokamak neutron source. Even with this reduction in effective neutron source level, the transmutation rates remain very interesting for transmutation reactors.

Chapter 21

Status of fusion development vis-à-vis a neutron source for FFH

Substantial progress has been made in recent years in achieving the plasma conditions required for a tokamak fusion neutron source. Using DT fuel, fusion powers exceeding 10 MW have been produced in both TFTR and JET. DT plasmas in these devices have approached the conditions required for $Q_p = 1$. Operating with deuterium plasmas, JT60-U has reached parameters exceeding those required for $Q_p = 1$ in DT plasmas.

These recent experimental accomplishments reflect a rapid growth in the understanding of the physics of anomalous transport in thermonuclear plasmas and of other aspects of tokamak behavior. Coupled with an extensive tokamak physics database gathered from dozens of tokamaks around the world [6], there now exists a knowledge base sufficient to design tokamak reactors or neutron sources that will achieve $Q_p \gg 1$, with high confidence.

Several concepts have been proposed for machines capable of reaching $Q_p \gg 1$ or even ignition, where the power produced by alpha particles in DT fusion reactions balances all energy losses from the plasma. While economic considerations dictate that electricity-producing fusion reactors operate with very large energy gain $Q_p \gg 10$, fusion neutron source applications such as spent nuclear fuel transmutation require only modest Q_p in order to be competitive. As we saw in the last section, $Q_p \approx 2 - 5$ is needed for a fusion neutron source coupled to a transmutation reactor with $k = 0.9$, and even smaller values of Q_p could be used with larger values of k. An upgrade of the JET facility could achieve $Q_p \sim 2$ within a few years. From the plasma performance standpoint, fusion has demonstrated a level of performance where fusion neutron source applications can be seriously considered for the immediate future. The remaining physics development required is associated with achieving higher annual neutron fluence accumulation, which involves a combination of achieving much longer burn pulses and higher reliability/availability.

The classical tokamak is a pulsed device with the current driven inductively. Although inductive burn pulses of 1000 s or more can be envisioned in a reactor-size $(R > 5 - 6 \text{ m})$ tokamak, the plasma discharge must ultimately be terminated while the inductive coil is recharged. High fluence accumulation applications requiring very long pulse or nearly steady-state operation, such as may be the case for transmutation of spent fuel, require that inductive current drive be supplemented, if not entirely replaced, by some form of non-inductive current drive.

Non-inductive methods for driving current in tokamaks on the basis of injection of energetic particles or RF waves are well developed. Since such methods are less efficient than inductive current drive, a central issue is the achievable gain Q_p. By optimizing the conditions favorable to the generation of the bootstrap current driven by pressure gradients, the current required to be driven by RF or beam sources can be minimized, greatly enhancing the possibility of achieving $Q_p \gg 1$. Significant progress has been made in developing well-confined tokamak operational regimes with high bootstrap current fraction. Extending these regimes to true steady state provides the focus for much of the present-day tokamak research. The status of resolving this most important (availability relies upon it) remaining physics issue is summarized in [28].

The successor to the present generation of large tokamaks, ITER [35], has been built through an international collaboration supported by the governments of the United States, Europe, Japan, the Russian Federation, South Korea, China, and India. The present ITER design is for a device that can operate in DT with an energy gain $Q_p \geq 10$ in the inductive mode and $Q_p \sim 5$ in the non-inductive or steady-state mode. The inductive mode is based on the type of tokamak operation that resulted in the high-performance regimes obtained in JET and JT- 60U. Non-inductive operation is based on the steady-state operating modes that are now in an advanced state of development. Total fusion power in either case will be $\sim$500 MW and the neutron wall loading will be 0.5 MW/m^2. These ITER plasma performance parameters are considerably more demanding than those identified in the previous section for a tokamak fusion neutron source for spent nuclear fuel transmutation (fusion power $\approx 10 - 100$ MW, neutron wall loading $\approx 0.1 - 0.5$ MW/m^2).

A substantial $\left(7.5 \times 10^8 \text{ US dollars}\right)$ tokamak technology R&D program for ITER has been carried out in parallel with the design effort. All technologies required for steady-state operation of a burning DT plasma were developed and tested in (near full scale for ITER, larger than full scale for a neutron source). Test facilities, including superconducting magnets, plasma heating, DT fuel processing, vacuum vessel fabrication, remote maintenance, and both plasma and nuclear heat removal systems, were developed and operated. Successful validation of these technologies has provided a high degree of confidence that a machine in the ITER class (and a smaller neutron source) can be built and reach its design performance parameters.

The timescale for construction of ITER has been about 10 years, and the capital cost is expected to be in the range of $(3 - 4) \times 10^9$ US dollars. This device demonstrates the integration in a single facility of the critical fusion physics and technologies required for a tokamak fusion neutron source for the spent nuclear fuel transmutation application

Existing austenitic steels which could be used for the first wall between the plasma and the surrounding material are estimated to have a lifetime against material damage by 14-MeV neutrons of 1–3 MW year/m^2, and ferritic steels and vanadium alloys which are under development for this application may have a considerably longer lifetime. Our estimate of the minimum first wall 14-MeV neutron annual fluence needed from a fusion neutron source for a subcritical transmutation reactor is ≤ 0.5 MW FPY/m^2, which is significantly lower than the annual fluence requirements ($\sim$1–2 MW a/m^2) which are usually projected for an electric power demonstration tokamak reaction [25]. However, there will be a significant fluence of fission neutrons on the first wall, as well as the 14-MeV fusion neutrons. Nevertheless, it would seem that the existing austenitic stainless steel should be adequate for the structural material in a fusion neutron source for a transmutation reactor, and it does not appear that a materials development program for advanced first wall materials would be required in support of a fusion neutron source for a transmutation reactor.

The nuclear (fuel, coolant, separation) technology being evaluated and developed in the ongoing ATW and OECD/NEA activities must be adapted to provide for tritium self-sufficiency of the fusion neutron source and to accommodate the tokamak neutron source geometry. For Pb–Bi coolant, one of the leading candidates under consideration, the addition of Li would seem to be a relatively straightforward adaptation, although the safety implications remain to be examined. Moreover, MHD effects for a liquid metal in a magnetic field are an additional issue that must be resolved. In any case, the requirement is to adapt technology otherwise being developed in the ATW nuclear program, not to take on the entire development of such technology.

It should be noted that a more robust superconducting magnet technology, such as is being investigated by researchers associated with MIT, would significantly enhance the prospects and performance characteristics of superconducting tokamak transmutation reactors.

Chapter 22

Fusion neutron enhancement of a breeding nuclear fission fuel cycle

As discussed previously, the availability of neutrons ultimately limits the amount of the nuclear energy in the uranium and thorium ores that can be extracted in critical nuclear fission reactors operating on "breeding" fuel cycles. More neutrons are needed to "breed" fissionable $^{239}Pu_{94}$, $^{241}Pu_{94}$, and higher fissionable transuranics by neutron capture in the majority isotope in uranium ore, $^{238}U_{92}$ (following the neutron transmutation/decay chain of Figures 3.1 and 3.6), and at the same time more neutrons are needed to fission enough fissionable atoms to maintain the fission neutron chain reaction in larger reactors.

Similarly, neutrons are required to produce fissionable $^{233}U_{92}$ and $^{235}U_{92}$ atoms by neutron capture in thorium $^{232}Th_{90}$ (followed by the neutron transmutation/ decay chains in Figure 3.6) and also to neutron fission enough atoms to maintain the power balance.

These two simultaneous requirements—(1) to maintain critical neutron chain reactions based solely on fission neutrons and (2) to also provide fission neutrons to breed fissionable plutonium from $^{238}U_{92}$ (as shown in Figure 3.6)—place practical limits on the amount of the potential nuclear energy that can be extracted from the uranium and thorium ores. These limits could be extended greatly (1–2 orders of magnitude) by the addition of an external neutron source to facilitate a "breeding" nuclear fuel cycle that created more fissionable plutonium from the second chain in Figure 3.6, and then fissioned it to extract energy.

Neutron sources that could be incorporated into nuclear (fission) reactors are on the horizon. Accelerator spallation neutron sources have been developed that could be used in nuclear (fission) reactors (e.g., [15]), and this application is being investigated also in Europe and Japan. However, the spallation neutron source is inherently a "point" source, implying that the source neutrons would not be broadly distributed spatially in a nuclear (fission) reactor but rather would be concentrated about the immediate location of the spallation source, further implying material radiation damage problems near the source location.

Nuclear fusion, on the other hand, could provide a strong and broadly distributed source of neutrons (e.g., [9–14, 23–25]). While combining a fission reactor with a fusion reactor neutron source unquestionably complicates the facility design, it provides the additional neutrons that are required to fully exploit the uranium and thorium fuel resources by enabling a "breeding" fuel cycle, as well as enabling a "burning" fuel cycle to recover the potential nuclear energy remaining in spent nuclear fuel removed from conventional fission reactors. There are also safety advantages to subcritical reactors, which have a much larger margin of error to the dangerous prompt super-critical condition than do conventional critical nuclear reactors. Fusion–fission hybrid reactors can be shut down to the decay heat level in seconds by simply switching off the electrical power supply to the fusion neutron source.

The ITER fusion experimental power reactor [35], which will begin operation in this decade, will provide 400 MWth (1.4×10^{20} neutrons/s) distributed over a broad area. The incorporation of a fast-spectrum (fission) nuclear reactor such as shown in Figure 6.4 (based on the metal-fuel Na-cooled fast reactor technology [18] that was developed on EBR-II and is being further developed for the new VTR fast reactor) within the ITER tokamak reactor magnetic configuration appears to be feasible [26,27]. China [40] and Russia [41,42] both have large-scale development programs for similar types of "fusion–fission hybrid" physics and technology.

The availability of a distributed neutron source of variable strength, such as the proposed SABR [26] D–T fusion tokamak neutron source extrapolated from ITER, could enhance nuclear (fission) power production by enabling "breeding" fuel cycles that transmute non-fissionable $^{238}U_{92}$ into fissionable $^{239}Pu_{94}$, $^{241}Pu_{94}$, and higher fissionable transuranics, and fission them (Figure 22.1 and Table 22.1).

Present (critical) nuclear reactors operate with a self-sustaining fission neutron chain reaction, which means that (on average) of the two to three neutrons released in a fission event exactly one neutron must cause another fission event. Creating and maintaining the conditions necessary for a critical nuclear reactor that can adjust to changes in composition as the fuel is fissioned and to changes in operating power requirements is a challenge that has been met by enriching the fuel in the fissionable isotope $^{235}U_{92}$ from the 0.72% in natural uranium ore to several percent or more in order to be able to maintain a self-sustaining fission chain reaction on fission neutrons alone for a sufficient length of operational time, and by incorporating additional fissionable material and compensating neutron-absorbing material, that can be adjusted as required to maintain criticality.

Maintaining the required fission reaction rate with fission neutrons supplemented by external fusion source neutrons would provide a great deal more flexibility [9–12]. This possibly would allow the use of natural uranium fuel without pre-enrichment, because the neutron breeding of fissionable plutonium from $^{238}U_{92}$ is an internal "self-enrichment" process associated with the transmutation of the dominant $^{238}U_{92}$ isotope in natural uranium, which is almost non-fissionable, into the very fissionable $^{239}Pu_{94}$ and $^{241}Pu_{94}$ and heavier fissionable transuranic isotopes via a neutron capture and radioactive decay transmutation chain (the second chain shown in Figure 3.6). Thus, the use of a variable strength external fusion neutron source could in principle create the possibility of eliminating or greatly simplifying the enrichment step, as well as facilitating the "breeding" fuel cycle. A broad fuel cycle system study [16] including the use of an undefined external neutron source with natural uranium fuel in this fashion

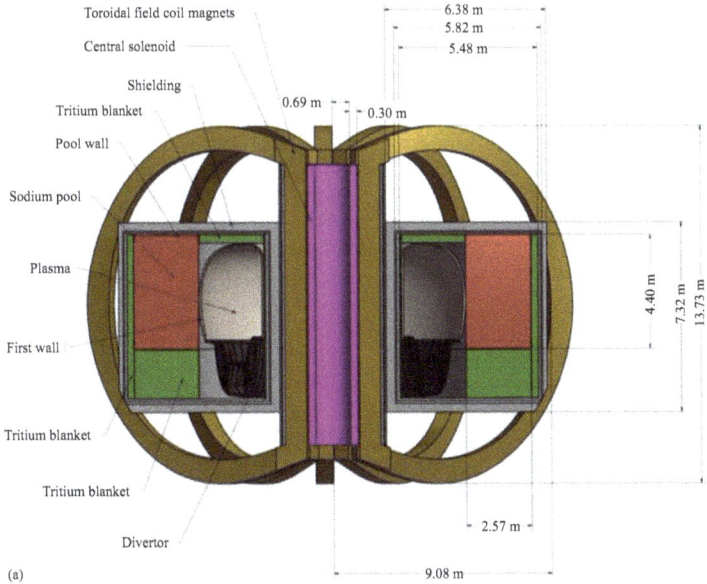

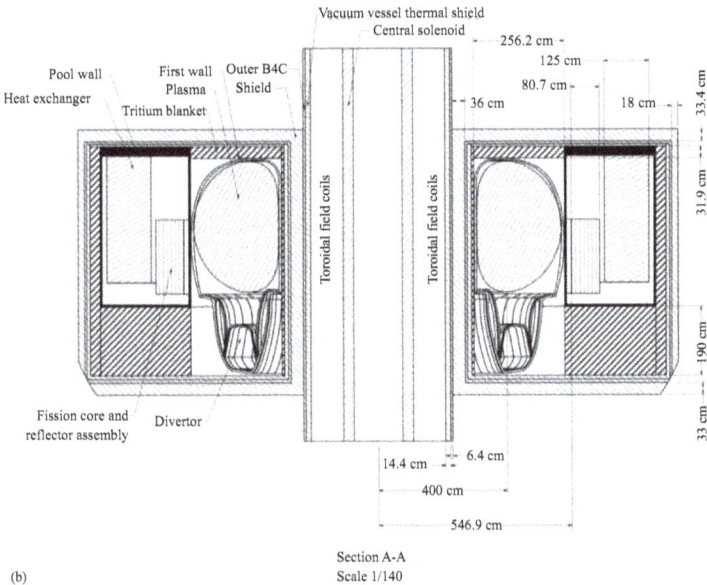

Figure 22.1 *SABR#2 fusion–fission advanced burner reactor [26] (The fission reactors are located in ten removable "sodium pools" (red) in an annulus just outboard of the tokamak fusion plasma neutron source.)*

Table 22.1   Major parameters and materials of the SABR#2 design [26]

Fast reactor core	
TRU fuel composition	40Zr-10Am-10Np-40Pu
BOL TRU mass	15,104 kg
BOL k_{eff}	0.973
Specific power	198.6 W/g HM
Fuel assembly	800
Fuel pin	469 per assembly, 375,200 total
Power density	256 kW/ℓ
Linear fuel pin power	12.3 kW/m
Sodium coolant mass flow rate	16,690 kg/s
Coolant temperature (T_{incool}, $T_{outcool}$)	628 K; 769 K
Fuel and clad temperature ($T_{maxfuel}$, $T_{maxclad}$)	1,014 K; 814 K
Clad and structure	ODS MA957
Electric insulator	SiC
Fuel/clad/bond/insulator/duct/coolant/wire (vol%)	22.3/17.6/7.4/6.5/9.3/35.3/1.5%
Tritium blanket	
Tritium breeder	Li_4SiO_4
Sodium coolant mass flow rate	1 kg/s (3.8 m/s)
Minimum and maximum blanket temperatures	450°C, 574°C
BOL TBR	1.12
Startup T required	0.7 kg
Reflector	
Materials—reflector assembly in-core (vol%)	ODS steel (58.1%), SiC (6.6%), Na (35.3%)
Materials—graphite reflectors (vol%)	Graphite (90%), Na (10%)
Shield	
Materials	Graphite, tungsten carbide, boron carbide, Na
Plasma	
Major radius	4.0 m
Plasma radius	1.2 m
Elongation	1.5
Toroidal magnetic field (on-axis)	5.6 T
Plasma current	10 MA
Inductive current startup	6.0 MA
Non-inductive current drive	4.5 MA
Bootstrap current fraction	0.55
H&CD power	110 MW (70 EC, 40 LH)
Confinement factor, H_{98}	1.2
Normalized β_N	3.2%
Safety factor at 95% flux surface	3.0
Maximum and BOL fusion power	500 MW and 233 MW
Maximum fusion neutron source strength	1.8×10^{20} n/s
Fusion gain ($Q_P = P_{fusional}/P_{extheat}$)	4.6

Table 22.1   (Continued)

Superconducting magnets	
Central solenoid	Adapted from ITER CS system
Maximum field	13.5 T
Flux core radius	0.69 m
CS coil thickness	0.3 m
Inductive flux	59 V s
TF coils	ITER TF system
Number	10 (18 on ITER)
Maximum field	11.8 T
TF bore dimensions	16.5 m (height) × 9 m (width)
PF ring coils	ITER PF system
Plasma H&CD system	
EC current drive efficiency	0.025 A/W
EC power	70 MW
EC upper quadrant modules (7.25 MW each)	4 (29 MW total)
EC equatorial modules (20 MW each)	2 (40 MW total)
LH current drive efficiency	0.035 A/W
LH power	40 MW
LH number of modules (20 MW each)	2
Divertor	
Materials	Tungsten, CuCrZr, Na cooled
Power to divertor[a]	105 MW; 31.5 MW
Heat flux	1–8 MW/m^2
Sodium coolant mass flow rate	0.09 kg/s
First wall	
Materials	Be, CuCrZr, ODS steel
Thickness	8.1 cm (1 cm Be, 2.2 cm CuCrZr, 4.9 cm ODS steel)
14-Mev fusion neutron power flux (average)	1.7 MW/m^2
Fission + scattered fusion neutron power flux	0.6 MW/m^2
Surface heat flux[a]	0.44 MW/m^2; 0.74 MW/m^2
Sodium coolant mass flow rate	0.057 kg/s
Sodium pool	
Number of modular pools	10
Mass of fuel per pool	1510.4 kg
Mass of Na per pool	22,067 kg
Power per pool	300 MW
Mass flow rate per pool	1,669 kg/s
Number of pumps per pool	2
Pumping power per pool	20 MW

[a]The first number is for a 50/50 split between charged particles and radiation across the separatrix and the second number is for a 15/85 split.

found that such source-driven systems compared favorably with more conventional critical reactor fuel cycles.

Several authors (e.g., [9–14) have examined Hans Bethe's suggestion to use a fusion neutron source to support a "subcritical" nuclear fission reactor for the purpose of neutron capture in the dominant $^{238}U_{92}$ isotope to produce ("breed") the very fissionable $^{239}Pu_{94}$ and $^{241}Pu_{94}$ as fuel for other nuclear fission reactors (i.e., a "breeding" fuel cycle). It is clear that such "fusion–fission hybrid breeder reactors" could also play a major role by providing fissionable $^{239}Pu_{94}$ to fuel other nuclear fission reactors or the reactor in question. The possibility of breeding fissionable fuel in nuclear reactors driven by either a tokamak fusion neutron source (e.g., [9,12]) or a mirror fusion neutron source (e.g., [10,13]) have been examined in some detail, and appear to be technically feasible.

22.1   Closing the fission fuel cycle with a burner FFH

In addition to enabling "breeding" fuel cycles to produce fissionable fuel from $^{238}U_{92}$, fusion neutron sources might play a more immediate role in the near term of nuclear power by "burning" (fissioning) the fissionable transuranics remaining in spent nuclear fuel (SNF) rather than burying them in high-level waste repositories (HLWRs), thus facilitating the "reprocessing" step in the nuclear fission fuel cycle shown in Figure 4.1. Fusion–fission hybrid reactors could be used to support closing the "back end" of the nuclear fission reactor fuel cycle by fissioning the transuranics (TRU) remaining unfissioned in the fuel discharged from present nuclear fission reactors [25–27,45]. The so-called "spent nuclear fuel" (SNF) is something of a misnomer because the TRU content of the fuel discharged from present reactors is not only highly radioactive but also highly fissionable in a fast reactor neutron spectrum, but not in the thermal neutron spectrum of PWR nuclear reactors (see Figure 6.3).

Calculations have been made for the SABR subcritical advanced burner reactor [35] shown in Figure 15.1, consisting of a tokamak plasma of ITER-like [32] parameters surrounded by ITER-like fusion technology and by a ring of sodium pools containing IFR-VTR type [17] metal-fuel (reprocessed TRU from LWR SNF) fast reactor modules (shown in Figure 6.4), to produce a combined fission + fusion power of 1,000 MWe. The fuel cycle calculations indicate that a 1,000-MWe SABR could fission the TRU in the spent fuel discharged annually from three 1,000 MWe PWRs, thereby increasing the total nuclear fission energy output of a given amount of uranium ore by 33%, while reducing the amount of spent fuel requiring storage in HLWRs by a factor of 10–100 [39,40]. Design basis accident safety analyses [46,49] indicate that, in the event of a serious malfunction, the power level in such a reactor could be reduced in a few seconds to the decay heat level by simply turning off the electric power to the plasma heating system, providing an extra degree of safety. These are discussed in Chapter 23.

The status of world-wide research on fission–fusion hybrids is substantial, particularly in Russia and China.

Chapter 23

Using fusion neutrons to achieve a burning fission fuel cycle

23.1 Introduction

At the present rate of nuclear power generation in the United States, enough spent fuel will soon have accumulated to fill a Yucca Mountain–type high-level waste repository (HLWR). The forecast for increased power generation by nuclear power in the next 30 years and over the coming century magnifies the issue of spent nuclear fuel (SNF) disposal. Between 2007 and 2010, the US Nuclear Regulatory Commission (NRC) accepted applications for 26 new light water reactors (LWRs), and the NRC expected applications for another five reactors in 2011 (Ref. [3]). These 31 reactors would increase the current nuclear power output of the United States by 30%, increasing the amount of discharge fuel needed to be stored in geological repositories by a similar amount.

Until very recently, the reference US option for disposal of SNF was (a) initially on-site storage and (b) shipping the fuel to a geological repository where it would be interred forever. With this predicted increase in nuclear power, a new geological repository of the same size as Yucca Mountain would be needed every 40–50 years.

A second option for spent fuel disposal is to introduce a multistrata fuel cycle in which following (a) initial on-site storage, (b) the spent fuel from LWRs is recycled in advanced burner reactor systems to burn essentially all the long-lived actinides that determine the requirement to demonstrate performance for a 100,000-year interment, and (c) then only the fission products and trace amounts of actinides are sent to long-term HLWRs. The multistrata fuel cycle would not replace geological repositories but would significantly reduce the number of HLWRs that are necessary.

Such multistrata fuel cycles have been widely studied for critical advanced burner reactors and for subcritical reactors supported by an external source of neutrons provided by an accelerator target embedded in the core and examined to a lesser extent for subcritical reactors supported by fusion neutron sources [11,15,16].

Utilizing subcritical reactors with a variable-strength fusion neutron source removes the criticality constraint on fuel residence time and allows for the fuel to remain in the reactor until the radiation damage limit is reached, which should result in fewer reprocessing steps and ultimately fewer repositories than would be needed in a multistrata fuel cycle utilizing critical burner reactors. The subcritical operation also provides an extra margin of safety to prompt critical, which should allow the burner reactor to be fueled with 100% transuranics (TRUs), instead of ~20% TRUs for critical reactors, which should result in fewer subcritical than critical burner reactors being necessary to support a given fleet of LWRs. On the other hand, the subcritical reactor with a fusion or accelerator neutron source would be more complex and expensive than a comparable critical reactor.

A fission–fusion hybrid is a subcritical fission reactor supplemented by a fusion neutron source. The fusion neutron source is chosen for three reasons. First, fusion is one of only two realistic existing options for a copious neutron source; second, the fusion neutron source strength is variable and can be increased or decreased to maintain a desired fission or thermal power level independent of the changes in reactivity throughout the cycle; and third, the fusion neutron source is a distributed neutron source capable of irradiating a larger volume of fuel than a point neutron source such as an accelerator target, the other option for such a large neutron source [15].

The following discussion focuses on various transmutation fuel cycles in the subcritical advanced burner reactor [12] (SABR), a fission–fusion hybrid reactor with a fusion neutron source based on ITER physics and technology [35] combined with a fast burner reactor based on the leading sodium-cooled fast reactor technology. Fuel residence time is limited in these fuel cycles by the radiation damage to the structural materials. This study examines three different fuel types— (a) TRUs (transuranics) from LWRs in a metallic fuel, (b) minor actinide (MA)-rich TRUs from which some of the Pu has been removed in a metallic fuel, and (c) MA-rich TRUs in an oxide fuel—for transmutation in the SABR.

23.2   The SABR tokamak FFH design concept

23.2.1   Overview

Figure 23.1 is a simplified three-dimensional model of the SABR tokamak design concept. The toroidal fusion plasma is surrounded on its outboard side by an annular

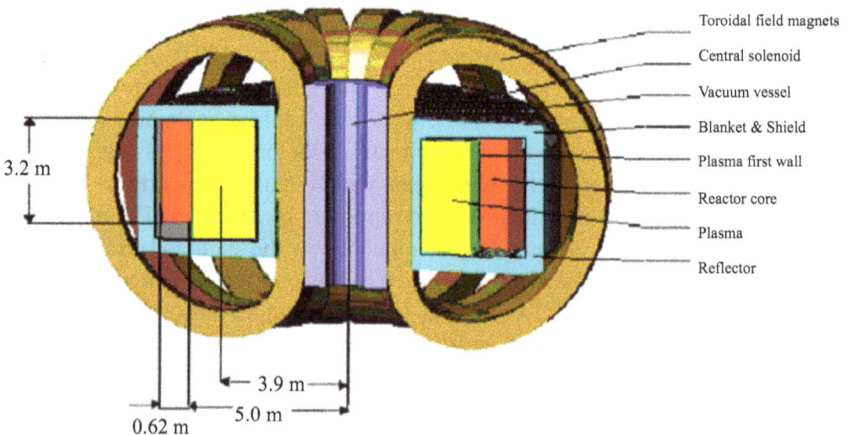

Figure 23.1   Three-dimensional model of the SABR concept

subcritical TRU-fueled fission core. The fission core consists of four concentric annular rings formed by vertical hexagonal fuel assemblies, surrounding the plasma of the fusion neutron source. The active core is 0.64 m thick by 2.0 m in active height (plus a 1-m fission gas plenum) and produces 3,000 MWth of thermal power (from fission, from thermal energy deposited by the fusion neutrons, and from exoergic nuclear reactions).

The annular core geometry is required by the neutron source geometry, but it may also have some advantages, such as negative sodium void and fuel expansion reactivity coefficients. Moreover, an annular core does not seem to have any disadvantages relative to a cylindrical core—for the same fuel and power density, annular and cylindrical cores with equal transmutation (fission) rates would have the same volumes.

Slightly different SABR core designs have been developed for three different fuel types. Two designs of metal fuel cores based on integral fast reactor (IFR) technology [17] were developed, one based on the TRU-Zr metal fuel under development at Argonne National Laboratory [17] and the other based on a TRU-MgO metallic fuel with the reduced plutonium composition specified in European fuel cycle scenario studies. The third fuel design is a TRU-MgO oxide fuel based on the same European fuel composition.

The Argonne TRU-Zr fuel composition is representative of the spent fuel discharged from LWRs, while the European TRU-MgO fuel composition is representative of an MA-rich fuel that would result after setting aside some of the plutonium recovered from LWR spent fuel for future use in starting up fast reactors.

The SABR reactor configuration of Figure 23.1 was designed to fit within the ITER magnetic configuration. The plasma and core region are surrounded by a 15-cm lithium oxysilicate blanket for tritium production, followed by a steel reflector and a multilayered shield to capture neutrons and gamma rays and to protect the toroidal field magnets from radiation damage.

A detailed geometric cross section of SABR, illustrating the locations of these various regions, is shown in the R-Z neutronics computation model of SABR shown in Figure 23.2. As a point of reference, SABR, with a plasma major radius <4 m, is about half the size (by volume) of ITER (Ref. [35]).

23.2.2 Fuel element and fuel assembly design

Originally, SABR was designed for the TRU-Zr metal fuel being developed by Argonne National Laboratory. The fuel is composed of 40Zr–40Pu–10Np–10Am by weight percent. The isotopic composition of the fuel is given in Table 23.1. The metallic fuel form was chosen because it has a high thermal conductivity, has the possibility of achieving the inherent safety characteristics of the IFR, Ref. [17], and has a fuel cycle in which the plutonium is never separated from the higher actinides. The fuel pins in SABR are based on a standard IFR-type metallic fuel pin design but coated with an electrical insulator to reduce the magnetohydrodynamic pressure drops associated with sodium coolant in a magnetic field.

(The SABR design power density was significantly lower than the usual IFR designs, and the effect of the insulator on thermal performance was taken into account.) The SABR fuel pins based on the composition in Table 23.1 are shown in Figures 23.3

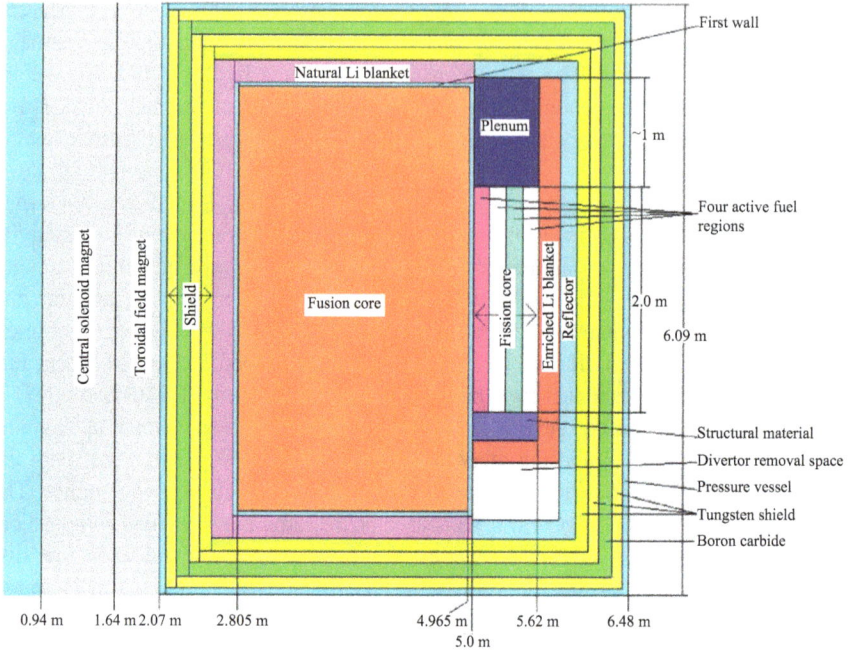

Figure 23.2   SABR#1 R-Z model

Table 23.1   Argonne TRU fuel composition [12]

Isotope	Mass percent at BOL
Neptunium-237	17.0
Plutonium-238	1.4
Plutonium-239	38.8
Plutonium-240	17.3
Plutonium-241	6.5
Plutonium-242	2.6
Americium-241	13.6
Americium-243	2.8

and 23.4. The basic SABR fuel assembly with 271 of these fuel pins is shown in Figure 23.5. The core consists of 918 such fuel assemblies arranged in four annular rings.

23.2.3   SABR fuel assemblies for European MA-rich fuel

The Karlsruhe Institute of Technology is examining a fuel cycle scenario of interest to some European countries, in which some of the plutonium is removed from spent fuel and set aside for future use, leaving the MA-rich fuel type as shown in

Table 23.2. This fuel composition was selected to have a minimal reactivity change with burnup to accommodate a subcritical burner reactor with an accelerator neutron source. The fuel consists of 45.7% plutonium and 54.3% MAs in a magnesium oxide matrix.

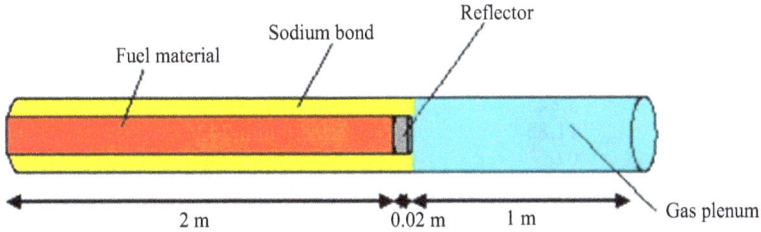

Figure 23.3   Axial view of SABR fuel pin (not to scale)

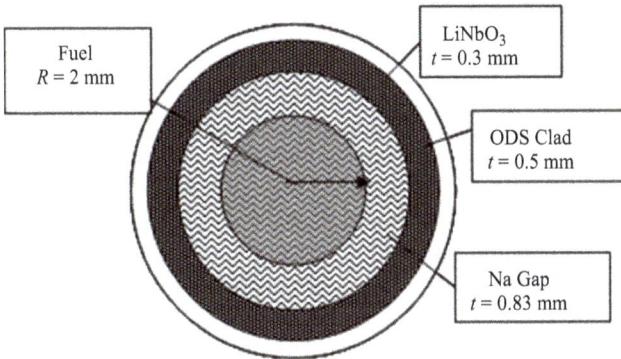

Figure 23.4   Metal-TRU fuel pin; ODS, oxide dispersion strengthened

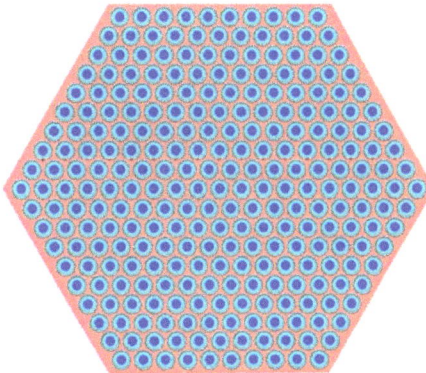

Figure 23.5   Metallic fuel assembly (15.5 cm across flats)

Table 23.2   European MA-rich fuel composition [24]

Plutonium vector		MA vector	
Isotope	**Mass percent**	**Isotope**	**Mass percent**
Plutonium-238	3.737	Neptunium-237	3.884
Plutonium-239	46.446	Neptunium-239	0.0
Plutonium-240	34.121	Americium-241	75.51
Plutonium-241	3.845	Americium-242*m*	0.254
Plutonium-242	11.85	Americium-242*f*	0.000003
Plutonium-243	0.0	Americium-243	16.054
Plutonium-244	0.001	Curium-242	0.0
		Curium-243	0.066
		Curium-244	3.001
		Curium-245	1.139
		Curium-246	0.089
		Curium-247	0.002
		Curium-248	0.0001

Two fuel types using the European fuel composition of Table 23.2 were examined. The first fuel type was a metal fuel. The same fuel pin and fuel assembly design described in Section 23.2.2 for the original SABR fuel were used, but with the fuel composition of Table 23.2 instead of that of Table 23.1 and an oxide matrix rather than a zirconium matrix.

An oxide fuel with the MA-rich composition of Table 23.2 was also considered. Since the oxide fuel has a lower heavy metal density than metallic fuel, a larger fuel volume is needed to operate in the same range of k_{eff} as the metal fuel designs. The SABR fuel pins and fuel assemblies were slightly redesigned to accommodate the oxide fuel. The European fuel pin design from the EFIT study [8] was used for the oxide fuel, resulting in the oxide fuel pin having a larger pin diameter and a smaller coolant-to-fuel volume ratio. This is possible because the oxide fuel has a much greater melting temperature than the metallic fuel (3,000 K for oxide and 1,350 K for metallic). The oxide fuel assembly has the same outer dimensions as the metallic fuel assembly but contains 217 fuel pins instead of 271. Each fuel pin with the oxide fuel has an outer pin diameter of 8.72 mm as compared to 7.36 mm for the metallic fuel. Figures 23.6 and 23.7 are representations of the oxide fuel pins and assembly, respectively.

A comparison of the major parameters for the oxide and the metal fuel pins and fuel assemblies is given in Table 23.3.

23.2.4   Fusion neutron source

Conservative ITER-like physics were adopted for the design of the SABR tokamak neutron source [13]. (By conservative, we mean that values of performance parameters that have already been achieved regularly in experiments were chosen for most physics design parameters, rather than the more favorable values anticipated

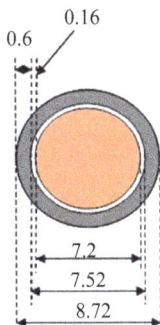

Figure 23.6   Oxide fuel pin [8]

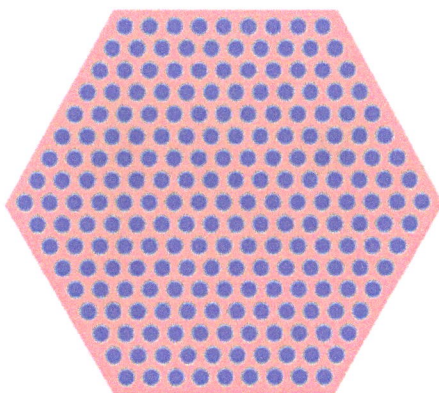

Figure 23.7   Oxide fuel assembly (~15.5 cm across flats)

in future experiments.) The neutron source was designed to produce a fusion power of P_{fus} 100–500 MW(thermal), which will be shown to be adequate to support the design objective of a total power in the fission core (from fission, fusion neutrons slowing down, and other exoergic reactions), of $P_{fis} = 3{,}000$ MW(thermal), under the range of subcritical operation envisioned.

The ITER fusion technological systems were adapted for SABR. The ITER single null divertor (not shown in Figure 23.1) and first wall were adapted for sodium coolant by scaling down to the SABR dimensions with the same coolant channel size. The ITER lower hybrid heating and current drive system was used to provide 100 MW of heating and to drive 7.5 mega-amperes of plasma current. The super-conducting magnet systems for SABR were scaled down [13] from the ITER design of magnets with a cable-in-conduit Nb_3Sn conductor surrounded by an INCOLOY® alloy 908 jacket and cooled by a central channel carrying supercooled helium, with maximum fields of 11.8 and 13.5 T, respectively, in the toroidal and poloidal field coils. The dimensions of the central solenoid coil were constrained by the requirement

Table 23.3   Key design parameters of metal and oxide fuel pins and assemblies

Parameter	Metal	Oxide	Parameter	Metal	Oxide
Length of fuel rods (m)	3.2	3.2	Total pins in core	248,778	199,206
Length of active fuel (m)	2	2	Diameter, flat to flat (cm)	15.5	15.5
Length of plenum (m)	1	1	Diameter, point to point (cm)	17.9	17.9
Length of reflector (m)	0.2	0.2	Length of side (cm)	8.95	8.95
Radius of fuel material (mm)	2	3.6	Fuel rod pitch (mm)	9.41	13.63
Thickness of clad (mm)	0.5	0.3	Pitch-to-diameter ratio	1.3	1.56
Thickness of Na gap (mm)	0.83	0.16	Total assemblies	918	918
Thickness of $LiNbO_3$ (mm)	0.3	0.3	Pins per assembly	271	217
Radius of rod with clad (mm)	3.63	4.36	Coolant flow area per assembly (cm^2)	96	108

to provide inductive start- up and to not exceed a maximum stress of 430 MPa set by matching ITER standards and INCOLOY properties. The dimensions of the 16 toroidal field coils were set by conserving tensile stress calculated as for ITER, taking advantage of an INCOLOY alloy 908 jacket for support.

It is intended that the fusion neutron source strength would be adjusted slowly (every week or so, perhaps) to compensate for the reactivity change due to fuel burnup to maintain a relatively constant fission core power level and temperature distribution within the reactor. There are many ways this could be accomplished, although a specific operational scenario has not yet been developed. The plasma power balance can be altered by changing the amount of heating power into the plasma, by changing the fueling rate, and by other changes to the operating conditions that affect the rate at which energy and particles escape from the plasma. The present design has 20-MW auxiliary heating units, which are too large to affect small changes in the neutron source level, so some incremental megawatt-level heating sources would probably have to be included. The fueling is by opening a valve and pumping gas into the chamber, and the amount of gas can be readily controlled. There are many possibilities for changing the rate at which energy and particles escape from the plasma. Thus, altering the plasma neutron source level would seem to be practically feasible and should have no impact on availability.

23.2.5   SABR fuel cycle

The SABR utilizes the out-to-in shuffling pattern depicted in Figure 23.8.

At the beginning of life (BOL), fresh fuel is placed in all four annular rings of the core. The fuel is irradiated for a burn cycle time and then shuffled inward by one ring, with the fuel in the innermost ring (ring 1 in Figure 23.8) being removed

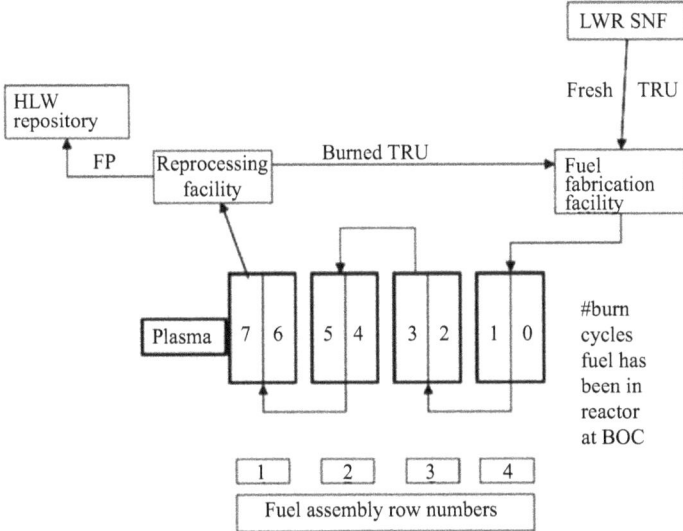

Figure 23.8   SABR out-to-in fuel shuffling scheme

from the core and sent to the reprocessing facility at the end of each burn cycle. Fresh fuel from the fabrication facility is loaded into the outermost ring, ring 4, and the fuel is irradiated for another burn cycle. This process is repeated, with the fuel composition fed into SABR soon reaching equilibrium.

In the reprocessing facility, the fission products are separated from the remaining TRUs (a conservative 99% separation efficiency is assumed). The fission products (and 1% of the TRUs) are sent to a HLWR. The remaining TRUs (and 1% of the fission products) are sent back to the fuel reprocessing facility where they are combined with fresh TRUs from LWRs and sent to the fuel fabrication facility, where new fuel elements and assemblies are manufactured and then placed back into the reactor.

Fuel shuffling and successive reloading within the tokamak toroidal field coil configuration of Figure 23.1 will be challenging and require some mechanical design ingenuity. Similar problems have been addressed by the ITER designers, who must provide for test assembly removal/insertion and for replacement of failed components in the same geometry. One can contemplate modular blocks of fuel assemblies, which are removed radially between toroidal field coils for refueling external to the coils, with blocks under the coils being rotated and then removed.

23.2.6   SABR fuel cycle simulations

A series of fuel cycle simulations was performed for the SABR transmutation system, to identify the characteristics of two types of fuel cycle that could be accommodated in SABR: (a) a TRU-burning fuel cycle in which all of the TRUs in the spent fuel discharged from LWRs were burned and (b) a MA-burning fuel cycle in which much of the plutonium from discharged LWR fuel was set aside for future use but all the MAs were burned. The length of the fuel cycle was determined by the radiation damage limit of the clad and structural material. Fuel cycles were examined with three different fuel types: metal-TRU, metal-MA, and oxide-MA.

Each fuel cycle simulation was evaluated based on multiple performance indicators: burnup, total TRU destruction, total plutonium destruction, total MA destruction, required fusion power, power peaking, LWR support ratio, radiation damage, decay heat to the repository, etc. The fuel cycle calculation was made with the ERANOS2.0 software package, using JEFF2.0 cross sections in 33 energy groups from 20 MeV down to 0.1 eV. A lattice cell calculation in P_1 transport theory and 1968 energy groups was performed on the fuel assembly, the energy groups were collapsed to 33 groups, and the assembly was then homogenized. The transport simulation in ERANOS was performed with BISTRO, a discrete ordinates transport solver. The flux solution was calculated using an S_8 quadrature set in the R-Z geometry of Figure 23.2, with 91 radial and 94 axial mesh points. Fuel depletion was simulated with the EVOLUTION module. EVOLUTION uses an average flux profile and depletes the fuel based on this profile for a given time period. To achieve an accurate isotopic burnup the fuel was depleted in 233-day time steps to account for the change in flux profile over time.

23.2.7   Accumulated radiation damage versus burnup for metal-TRU fuel

The first fuel cycle issue examined was the effect of clad radiation damage limit on fractional fuel burnup in a fuel residence time. Simulations were run for the SABR out-to-in shuffling pattern for irradiation times corresponding to 100, 200, and 300 displacements per atom (DPA), as well as for a hypothetical once-through fuel cycle with a radiation damage limit sufficient to achieve >90% burnup before the fuel is removed from the reactor. Radiation damage limits of 150–200 DPA are anticipated for clad and structural materials presently under development. The 300-DPA limit was investigated to determine if there is a strong incentive for developing new clad materials able to withstand a higher radiation damage dose. The once-through cycle was examined to determine what radiation damage limits would be needed to achieve high burnup of the TRU fuel without reprocessing and to examine the power distributions that would result in such a low reactivity core.

The simulations show that the relationship between radiation damage and burnup is linear in the regime from 100 to 300 DPA. This linear relationship results in linear increases in fusion power and TRUs burned per residence in this regime. The results are summarized in Table 23.4.

The TRU burnup rate depends on the fission rate, of course, and the fission rate decreased as the fusion rate increased to compensate for reactivity loss [recall that it is the total thermal power in the fission core that is held constant at 3,000 MW (thermal)]. This accounts for the downward trend in TRUs burned per year from the 100-DPA cycle to the once-through cycle. As can be seen from Table 23.4, the ratio of fission power to fusion power in the recycling fuel cycles varies from >30 at BOL to ~7.5 at the end of the equilibrium fuel cycle (EOC).

Table 23.4   Summary of radiation damage versus burnup

Parameter				
Cycle	100 DPA	200 DPA	300 DPA	Once through
Burn cycle length time (days)	350	700	1,000	4,550
Four-batch residence time (year)	3.83	7.67	10.95	49.8
Fission power @MW(thermal)#	3,000	3,000	3,000	3,000
FIMA (%)	16.7	23.8	31.6	87.2
Region power peaking BOC/EOC	1.7/1.8	1.8/2.0	1.8/2.0	2.0/2.1
BOL P_{fus} (MW)	73	73	73	73
BOC P_{fus} (MW)	155	240	286	1,012
EOC P_{fus} (MW)	218	370	461	1,602
BOL k_{eff}	0.972	0.972	0.972	0.972
BOC k_{eff}	0.940	0.894	0.887	0.784
EOC k_{eff}	0.916	0.868	0.834	0.581
TRUs burned/year (kg)	1,073	1,064	909	545
Support ratio (75% availability)	2.9	3.2	3.6	2.2
Clad damage (DPA)	97	214	294	1,537

Since a 1,000-MW (electric) LWR produces ~250 kg of TRUs/year, a support ratio of LWRs per SABR can be defined by dividing the SABR TRU destruction rate by the LWR production rate. For this purpose, we assume that SABR operates at 75% availability, thus taking into account refueling downtime and unscheduled downtime.

The assembly-average power peaking (the assembly-average power in the first ring to the core-average power over all four rings) indicated in Table 23.4 is generally <2.0. The power is relatively uniform, with peak-to-average factors of 2 or less, except for the once-through cycle where the very different composition of the fuel in adjacent rings produces large power peaking. The detailed power distribution is the most limiting for the 300-DPA case (Figure 23.9).

The distribution of accumulated fast-neutron (>100 keV) fluence and the DPA at the end of the 300-DPA equilibrium cycle are plotted in Figures 23.10 and 23.11, respectively.

The jumps in the distributions occur between rings of assemblies that have been in the reactor for different numbers of burn cycles. This sort of ring-to-ring power peaking can be handled by flow zoning among the rings of assemblies [17], and the within-assembly DPA gradient can be reduced by rotating the assemblies when they are shuffled between rings. No effort has been made yet to optimize the within-assembly power distribution.

The decay heat in the repository was calculated with ORIGEN-S. Fast-group cross sections were imported into ORIGEN-S, and the fuel was then depleted under a constant flux until the burnup reached the same level of burnup seen in ERANOS. The calculation of decay heat to the repository was done assuming reprocessing separation efficiency of 1%, meaning 99% of the fission products and 1% of the TRUs go to the repository on each reprocessing step.

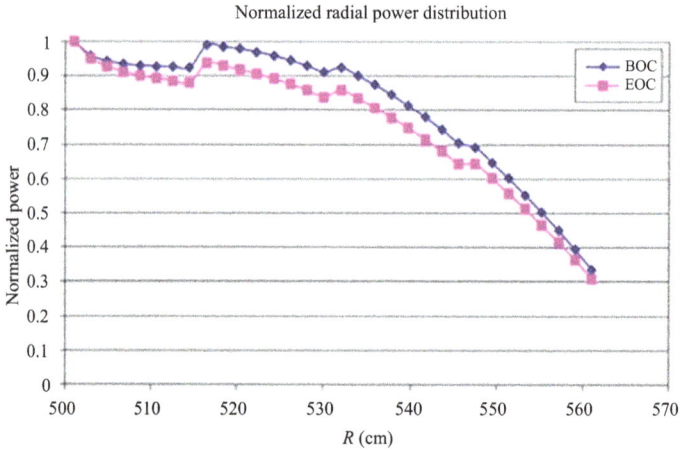

Figure 23.9   Power distribution for the TRU burner fuel cycle at end of the 300-DPA equilibrium cycle

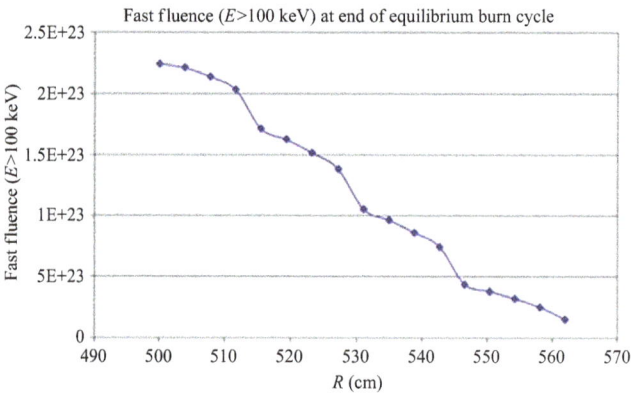

Figure 23.10   Fast-neutron fluence at end of the 300-DPA equilibrium fuel cycle

Figure 23.12 shows the decay heat to the repository for each of the four TRU burner fuel cycles, as well as for unprocessed fuel discharged after a typical LWR once-through fuel cycle. The decay heat to the repository is in proportion to the number of reprocessing steps, which varies inversely with the DPA limit. Clearly, a reduction in long-time decay heat to the repository of more than a factor of 10 could be accomplished by recycling the TRUs in LWR spent fuel in SABR. While the decay heat is inversely proportional to the DPA limit, Figure 23.12 indicates that a substantial factor of 10 reduction is achievable with a radiation damage limit of 100–200 DPA and it is not critical to increase the DPA limit further in order to make transmutation of TRUs realistic.

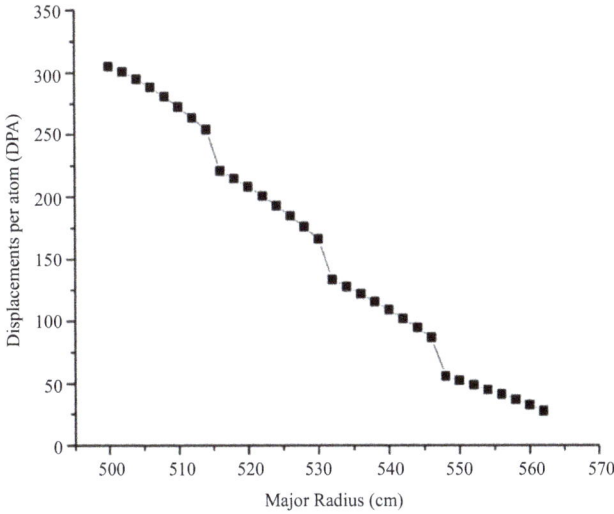

Figure 23.11 Displacements per atom at the end of the 300-DPA equilibrium fuel cycle

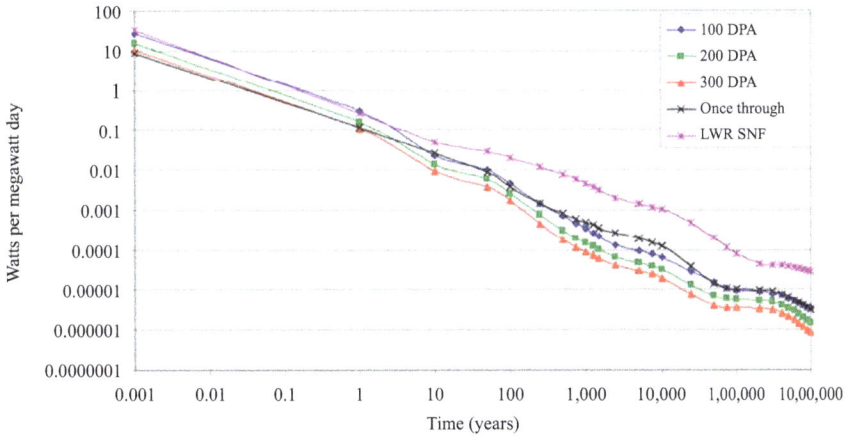

Figure 23.12 Decay heat to the high-level waste repository

The initial calculations for fuel residence versus radiation damage were done assuming a fuel smear density of 100% and no rotation of fuel assemblies with shuffling (i.e., the same face of the assembly would be located inboard as the assembly was shuffled from the outermost to the innermost ring over the fuel cycle). The calculations on the reference 200-DPA cycle were repeated to investigate the effect of (a) utilizing a smear density of 95% to accommodate fuel swelling and expansion and (b) rotating the fuel assemblies by 180° each time they

Table 23.5   Comparison of rotated and non-rotated 200-DPA fuel cycles with metal fuel

	200-DPA rotated (95% density[a])	200-DPA non-rotated (95% density)	200-DPA non-rotated (100% density)
Fission power [MW(thermal)]	3,000	3,000	3,000
BOL mass (kg HM)	30,254	30,254	31,846
FIMA (%)	25.6	25.6	24.1
Region power peaking BOC/EOC	1.7/1.9	1.7/1.9	1.8/2.0
BOL P_{fus} (MW)	172	172	73
BOC P_{fus} (MW)	302	317	240
EOC P_{fus} (MW)	401	429	370
BOL K_{eff}	0.945	0.945	0.972
BOC K_{eff}	0.878	0.863	0.894
EOC K_{eff}	0.831	0.817	0.868
Cycle reactivity change (pcm)	−6,441	−6,526	−3,351
TRUs burned/year (kg)	1,027	1,023	1,064
Support ratio (100%)	4.1	4.1	4.2
Support ratio (75%)	3.1	3.1	3.2
Clad damage (DPA)	212	218	214

[a]The high values of the smear density are perhaps unrealistic, so the heavy metal mass is the relevant fuel parameter. The actual design (pin and assembly) sizes would be somewhat different than the fuel design given in this paper in lower-smear-density designs.

were shuffled. The results are summarized in Table 23.5. The effect of rotating the assemblies is minimal; the regional power peaking and the radiation damage are reduced by 2% and 3%, respectively, while the rest of the parameters remain the same.

The effect of changing the fuel smear density (a proxy for the total heavy metal mass) has a significant impact on the fuel cycle. The major effect of lowered heavy metal mass is a reduction in multiplication constant k, with a corresponding increase in the required fusion power $P_{fus} = const(k/(1-k))P_{fis}$. The decrease in heavy metal mass also reduces the amount of waste to the repository and the amount of decay heat to the repository both by ~5%.

23.2.8   Minor actinide burner

The MA burner fuel cycle analysis emphasizes fissioning the MAs in spent fuel while setting aside the plutonium for other uses, as specified in the European studies of reactors to burn MAs. The same 200-DPA, four-batch-with-rotated-assembly fuel cycle was analyzed for both the MA-oxide and the MA-metallic fuel burner fuel cycles (see Table 23.6). The fuel cycles were evaluated based on the same criteria as used for the TRU burner fuel cycle.

The change in reactivity throughout the fuel cycle is greater in the oxide fuel because more plutonium is burned. This requires a greater change in fusion power

Table 23.6   Minor actinide burner fuel cycle comparisons

	Minor actinide–metal fuel	Minor actinide–oxide fuel
Fission power [MW(thermal)]	3,000	3,000
Four-batch residence time (year)	7.67	7.67
BOL mass (kg HM)	49,985	47,359
BOC mass (kg HM)	48,468	45,658
EOC mass (kg HM)	46,441	43,542
Delta mass (kg)	2,027	2,110
Loading outer (kg)	13,040	12,345
Heavy metal out (kg)	11,013	10,234
FIMA (%)	15.5	17.1
Region power peaking BOC/EOC	1.5/1.6	1.3/1.5
BOL P_{fus} (MW)	489	515
BOC P_{fus} (MW)	190	195
EOC P_{fus} (MW)	246	325
BOL K_{eff}	0.889	0.909
BOC K_{eff}	0.949	0.959
EOC K_{eff}	0.932	0.936
Cycle reactivity change (pcm)	−1,922	−2,552
TRUs burned/year (kg)	1,089	1,122
Minor actinides burned/year (kg)	853	674
Plutonium burned/year (kg)	236	469
Uranium generated/year (kg)	31	21
Ratio of decay heat to LWR	0.10	0.10
SNF decay heat at 100,000 years		
Support ratio (100%)	34.1	27.0
Support ratio (75%)	25.6	20.2
Clad damage (DPA)	203	201

from beginning of cycle (BOC) to EOC for the MA-oxide fuel. The fusion power required to maintain 3,000 MW of thermal power in the core varied from P_{fus} 200 to 500 MW in these fuel cycles, and the rate of MA fission (destruction) were 850 and 675 kg per effective full-power year for the metal form and the oxide form of the fast reactor fuel, respectively.

The TRU transmutation rate for the MA fuel is 1,089 kg/year for the metal fuel and 1,122 kg/year for the oxide fuel. The metal fuel burns more MAs than the oxide fuel; 78.3% of the TRUs burned in the metal fuel are MAs compared to 58.9% of the TRUs burned in the oxide fuel. This is because the metal fuel is in a harder spectrum, making the fission cross section of the MAs more competitive with the fission cross section of the plutonium in the system. The normalized flux spectra for the oxide fuel and the metal fuel are shown in Figure. 23.13.

The neutron transport calculation treats the fusion neutrons as a volume source in the plasma region just inboard of the annular subcritical multiplying fission core region, both surrounded by reflectors and shields. The fusion neutrons slow down, cause fissions and (n, 2n) reactions, which produce neutrons, which slow down and

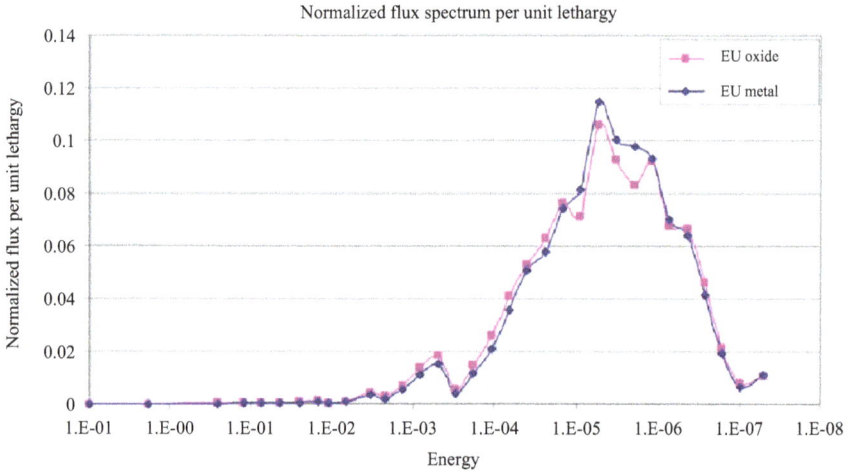

Figure 23.13   Normalized neutron flux spectra for MA-oxide and MA-metallic fuels

cause more fissions and (*n*, 2*n*) reactions, etc. The fusion source neutrons cause the spike in the group containing 14.1 MeV, which is more pronounced adjacent to the plasma source. The spectra shown in Figure 23.13 are for a location in the middle of the first ring adjacent to the plasma source.

The spectral differences in the metallic and the oxide fuel cores are dictated by the competition of two effects: the fuel-to-coolant ratio and the amount of matrix material. The metallic fuel, with ~0.35 cm^2 of coolant per pin, has a much larger fuel-to-coolant ratio than the oxide fuel, with ~0.69 cm^2 of coolant per pin, which tends to make the spectrum harder in the metal fuel. On the other hand, the metal fuel has ~60% matrix material, while the oxide fuel has ~45% matrix material, which tends to make the oxide spectrum harder. The harder spectrum for the metal fuel is a result of the metal fuel design having a tighter lattice and less coolant per fuel pin than the oxide fuel design. The larger percentage of matrix material in the metallic fuel than in the oxide fuel is more than compensated for by the smaller amount of coolant per pin for the oxide fuel design, resulting in a harder spectrum for the metallic fuel. Except immediately adjacent to the fusion neutron source, the neutron spectrum is more determined by the core composition than by the original energy of the neutron source (fission or fusion), so that the spectrum in SABR is very similar to what it would be in a critical IFR with the same core composition.

The LWR support ratio for the MA burner fuel cycle is defined as the ratio of MAs burned in SABR to the amount of MAs produced in a 1,000-MW(electric) LWR, typically ~25 kg of MAs/year. Note that this definition of support ratio for MA burning is different than the definition for TRU burning used previously. The LWR support ratios for the SABR metallic and European oxide fuels, assuming 75% availability, are 25.6 and 20.2, respectively, 1,000-MW(electric) LWRs supported by a single SABR.

The decay heat to the repository in this system is very similar for both the oxide fuel and the metallic fuel; the overall burnups are 17.1 and 15.5%, respectively, reducing the need for repository capacity by 8.08 and 7.91% for the oxide system and the metallic system, respectively, not accounting for the plutonium that has been set aside for further use in mixed oxide or fast reactor systems. The decay heat to the repository in both of these cases is shown in Figure 23.12, together with the heat that would be produced if the LWR spent fuel was just placed in the repository.

The rest of the evaluation criteria—power peaking, radiation damage, and overall TRU destruction rate for the metallic fuel and the oxide fuel in the MA burning cycle—were all very similar throughout the fuel cycle. This is a result of the fuels having similar BOL, BOC, and EOC reactivities and fusion powers. The oxide fuel performs better in terms of overall burnup, 17.1–15.5% for the oxide fuel and the metallic fuel, respectively. The MA fuels have very similar fuel cycle performances in SABR. The biggest difference is in the ratio of transmutation of MAs to the transmutation of plutonium. The choice of metallic- or oxide-fueled MA burners would be determined by this as well as by how each fuel performs in regard to safety in an accident scenario.

23.3   Conclusions

In order to impact climate change over the next century, a massive source of clean, carbon-free energy is needed. Nuclear power provides the only technically credible option. The major technical problem preventing the widespread expansion of nuclear power is spent nuclear fuel (SNF). The US government currently has no plan to deal with SNF generated by commercial power plants. Plant operators have been forced to stockpile SNF on-site. The common suggestion to deal with SNF is the high-level waste repository (HLWR). A repository could store the SNF for on the order of a million years until it was no longer radioactive. Various organizations have proposed several HLWRs, but none of them are yet close to becoming a reality in the United States.

It is clear from the results of this chapter that FFH burner reactors could reduce substantially the number of HLWRs required to resolve the high level waste problem of nuclear fission reactors.

Chapter 24

Fuel cycle methodology, summary and conclusions

We have investigated two types of fuel cycle for a SABR consisting of an annular, Na-cooled fast reactor surrounding a tokamak fusion neutron source. The first fuel cycle type is one in which all of the TRUs in LWR SNF are transmuted in a SABR, and the second fuel cycle type is one in which some of the plutonium in LWR spent fuel is set aside for future use and the remaining plutonium plus the MAs are transmuted in a SABR. In both fuel cycle types the fuel residence time between reprocessing steps was set by clad radiation damage limits (200 DPA reference value), and the separation of TRUs from fission products was assumed to be only 99% efficient. We found that, by repeated recycling of the TRU fuel discharged from SABR with a blend of fresh TRUs discharged from LWRs, the decay heat of the repository content could be reduced by a factor of ~10 at 100,000 years relative to the decay heat if the discharged fuel from LWRs was buried directly (Figure 24.1). Noting that decay heat load was the limiting design factor for Yucca Mountain capacity, this reduction in decay heat implies a corresponding reduction by a factor of 10 in HLWR capacity requirement. This result is based on the con-servative assumption that the actinide–fission product separation efficiency is only 99%. We note that there are other measures (e.g., Sr and Cs management and

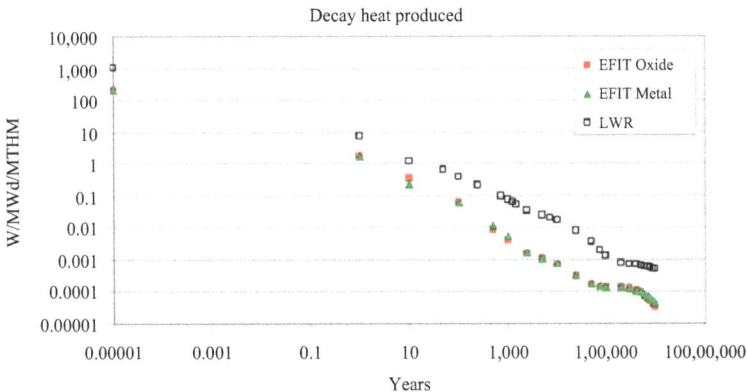

Figure 24.1   Decay heat produced in HLWR for MA burner oxide and metallic fuels

cooling before storage) for reducing the required repository capacity, and they are not incompatible with the transmutation solution proposed here. A 3,000-MW (thermal) SABR operating on such fuel cycles, with 75% availability, would be capable of burning all of the TRUs discharged annually from 3 LWRs of 1,000 MW (electric), or burning all of the MAs and some of the plutonium discharged from 20 to 25 LWRs of 1,000 MW (electric). Thus, one could envision a nuclear fleet with 75% of the energy produced by LWRs and 25% of the energy produced by SABRs that burned all of the TRUs discharged from the LWRs. Alternatively, one could envision a nuclear fleet with 95% of the energy produced by LWRs and 5% produced by SABRs that burned the MAs (primarily) and some of the plutonium discharged from LWRs, while plutonium was accumulated to start up fast reactors.

Breeder reactors take advantage of the high neutron-per-fission yield of fissile isotopes, particularly ^{239}Pu, in fast neutron spectra to supply extra neutrons beyond those necessary to sustain the fission chain reaction. These excess neutrons are captured in fertile material such that no more fissile material is produced than was consumed. Burner reactors leverage a fast neutron spectrum to transmute, preferably by fission, transuranic (TRU) isotopes that remain in the spent fuel discharged from thermal reactors. These TRUs, which constitute a substantial fraction (tens of percent) of burner reactor fuel, would otherwise be sent directly to a geological repository and dominate the long-term radiotoxicity and decay heat of the used nuclear fuel. Some reactor designs incorporate the aspects of both breeders and burners. They are intended to operate in an integrated fuel cycle, mixing their own discharged fuel with used fuel from other reactors and depleted uranium to form the next fuel loading. One of the most mature of these integrated reactor concepts is the integral fast reactor (IFR); many of its design decisions and material choices reflect its integrated fuel cycle and the very hard neutron spectrum which that cycle requires.

A subcritical advanced burner reactor (SABR) concept that addresses the waste problem is being developed at the Georgia Institute of Technology. SABR is a sodium-cooled, 3,000 MW(thermal) annular fast reactor consisting of four assembly rings surrounding a toroidal plasma. The fission core operates in the subcritical regime; the plasma supplies an external neutron source via the D–T fusion reaction. The fuel pins of SABR are loaded with TRUs processed from used fuel from light water reactors (LWRs) that is fissioned to a high atomic percent burnup. Neutrons leaking from the fission core are captured in surrounding tritium breeding blankets to produce fuel for the fusion reaction. Because of the subcritical operation, SABR is postulated to be able to be fueled with 100% TRU fuel discharge from LWRs, as contrasted with the tens of percent envisioned for critical reactors [24]. Fuel cycle calculations [14,15] indicate that SABR could consume TRUs at triple the rate that an LWR of the same power output produces them; a future reactor fleet might then produce 75% of its electricity in LWRs and 25% in SABRs and send no TRUs, other than losses from reprocessing, to repositories. SABR is based on existing technologies developed for the IFR and on ITER physics and technology and could be deployed by the midcentury.

Because the plasma and technology performance required for an economical fusion power plant significantly exceeds that which will be demonstrated in ITER, developing fission–fusion hybrid (FFH) reactors with ITER-level plasma and technology requirements in parallel with the further plasma and technology development needed for pure fusion power would allow for substantial accumulation of power reactor operating experience with tokamaks prior to the introduction of pure fusion power plants into reactor fleets. In the near term, these FFHs would likely be devoted to burning actinides, while in the longer term, using them for fissile production becomes more desirable as easy-to-extract ^{235}U becomes depleted. The SABR studies indicate the efficacy of the FFH in the burner role. An important question then arises: Could a similar hybrid reactor make a useful contribution as a breeder?

Ultimately, any FFH design must compare favorably to critical fast fission reactor designs to justify the added cost and complexity of including the fusion element. Such a comparison must include comparative analyses of (a) the fissile production capability of subcritical and critical fast breeder reactors, (b) the dynamic responses of the subcritical and critical designs to various accident scenarios, (c) the overall cost and reliability of the system of breeder reactors and reprocessing/refabrication facilities, and (d) the resistance to proliferation. The first step in this evaluation is to establish a realistic subcritical (FFH) fast breeder reactor technology for comparison with critical fast breeder reactor concepts.

The radial blanket fuel for SABR is a U-10Zr alloy at 85% smear density; it is similar to the Mark-I fuel pins for EBR-II, except with 10% Zr by weight instead of 5% fissium (a description of fissium is given elsewhere). The ^{235}U enrichment is 0.25% by weight. The 10% Zr was chosen because of its beneficial effect on the fuel melting temperature and on fuel-cladding interactions. While the lower uranium volume fraction likely has a slightly negative effect on breeding, the plasma-side edge of the inner radial blanket receives the fusion neutrons most directly and thus has a relatively high power for a fast reactor blanket; therefore, the thermal considerations are of primary importance. For that same reason, the pin diameter and number of pins per assembly are kept the same as in the driver fuel rather than using the fewer, larger pins found in most blanket assembly designs. The Mark-II fuel pins were limited to 3 at.% burnup in EBR-II due to burnup-induced swelling; the SABR pins are similarly limited. The axial blankets are the same composition as the radial blankets, but at 75% smear density to allow the fission gases from the driver fuel easy access to the plenum. Because of their lower smear density, the burnup limit of the axial blankets matches that of the driver fuel. The breeding of fissile material occurs by neutron capture in fertile isotopes within the fission annulus. The primary fertile isotope is ^{238}U, which captures a neutron and then decays by beta emission twice before becoming Pu239.

Isotope	Weight percent
238_{Pu}	2.102
239_{Pu}	58.258
240_{Pu}	25.976
241_{Pu}	9.76
242_{Pu}	3.904

$$\ce{^{238}_{94}Pu} + \ce{^{1}_{0}n} \rightarrow \ce{^{239}_{94}Pu}$$

and

$$\ce{^{240}_{94}Pu} + \ce{^{1}_{0}n} \rightarrow \ce{^{241}_{94}Pu}.$$

Because ^{238}U is the most common isotope of the three in the fresh driver fuel, most of the fissile production is through the first reaction, though all three substantially contribute. In the blanket assemblies, only ^{238}U is initially present, so nearly all fissile production that occurs there is via ^{239}Np.

24.1   Calculation methodology

Both the burner SABR and the breeder SABrR were modeled with the European Reactor ANalysis Optimized calculation System (ERANOS), a fast reactor code system developed to model the Phénix and SuperPhénix reactors. ERANOS employs the European Cell COde (ECCO) to collapse 1968-group JEFF2.0 cross sections within each reactor lattice cell to the 33 groups used in core calculations, ranging from 20 MeV down to 0.1 eV. The core geometry was described in R-Z cylindrical geometry and the core calculations performed in the ERANOS discrete ordinates transport module BISTRO using an S8 quadrature with 132 radial and 216 axial mesh points. The fuel was depleted for 100 days in each burnup step in the EVOLUTION module before reperforming the core neutron flux calculations.

At each depletion step, the neutron source multiplication k_{mult} is calculated, and the fusion neutron source strength is adjusted such that the fission annulus output is 3,000 MW(thermal). The fusion power P_{fus} required to maintain a given fission power P_{fis} is determined by k_{mult}, the average number of neutrons released per fission v, the energy released per fusion E_{fus}, and the energy released per fission E_{fis}:

$$P_{fus} = P_{fis}\left(\frac{1 - k_{mult}}{k_{mult}}\right) \tag{24.1}$$

It is important to note that k_{mult} differs from the more familiar k_{eff}. The TBR is also calculated at each step to determine if enough tritium is being produced to fuel sustained operation of the fusion neutron source. Practical experience indicates the TBR > 1.15 is generally adequate for tritium self-sufficiency.

TBR is defined as

$$\mathrm{TBR}(t) = \frac{\int_V \Sigma_c^{Li}\phi(\boldsymbol{r}, t)\mathrm{d}V}{\int_V S(\boldsymbol{r}, t)\mathrm{d}V}, \tag{24.2}$$

where S is the fusion neutron source.

This only accounts for production of T by ^{6}Li capture and thus is a conservative (low) estimate of the TBR, as T produced in the threshold reaction in ^{7}Li is not counted in the ERANOS calculation. However, since the tritium breeding material is highly enriched in ^{6}Li and the cross section for production via that route is much higher, the approximation should be quite close to the true tritium production rate.

FBR is the instantaneous ratio of the production rate of fissile atoms to their destruction rate, whether through fission or parasitic capture:

$$\text{FBR}(t) = \frac{P(t)}{D(t)} \tag{24.3}$$

The production rate is calculated by integrating the capture rates of the fertile isotopes over the reactor volume:

$$P(t) = \int_V \left(\Sigma_c^{238\text{U}}(r)\phi(r,t) + \Sigma_c^{238\text{Pu}}(r)\phi(r,t) + \Sigma_c^{240\text{U}}(r)\phi(r,t)\right)\mathrm{d}V. \tag{24.4}$$

Though ^{239}Np, rather than ^{238}U, is technically the precursor to ^{239}Pu, ^{239}Np exists in the reactor in a near steady state after its first few half-lives. Thus, by approximately day 20 of fuel residence time, the decay rate of ^{239}Np is equal to the capture rate of ^{238}U. The destruction rate is the volume-integrated absorption rate for all of the fissile isotopes. Only ^{235}U, ^{239}Pu, and ^{241}Pu exist in substantial amounts in the reactor, so other fissile isotopes are omitted from the summation:

$$D(t) = \int_V \left(\Sigma_c^{235\text{U}}(r)\phi(r,t) + \Sigma_{\text{abs}}^{239\text{Pu}}(r)\phi(r,t) + \Sigma_{\text{abs}}^{241\text{Pu}}(r)\phi(r,t)\right)\mathrm{d}V. \tag{24.5}$$

Substituting these expressions for the production and destruction rates into Eq. (24.3), we have

$$\begin{aligned}\text{FBR}(t) &= \frac{P(t)}{D(t)} \\[6pt] &= \frac{\int_V \left(\Sigma_c^{238\text{U}}(r)\phi(r,t) + \Sigma_c^{238\text{Pu}}(r)\phi(r,t) + \Sigma_c^{240\text{U}}(r)\phi(r,t)\right)\mathrm{d}V}{\int_V \left(\Sigma_c^{235\text{U}}(r)\phi(r,t) + \Sigma_{\text{abs}}^{239\text{Pu}}(r)\phi(r,t) + \Sigma_{\text{abs}}^{241\text{Pu}}(r)\phi(r,t)\right)}\end{aligned} \tag{24.6}$$

24.1.1   Design constraints

There were four hard constraints placed on the reactor design that, if violated, were a termination point for that particular case. First, the TBR must not fall below 1.15. This "practical" value was chosen because tritium self-sufficiency is a requirement for sustained fusion operation, and previous calculations indicate that this excess above unity allows for losses due to inefficiency in tritium collection and for the radioactive decay of any tritium in inventory throughout the operating and refueling cycles. Second, the radiation damage limit of the clad must not be exceeded. The damage limit of ODS MA957 in a fission spectrum is estimated at either 200 DPA or at an accumulated fast fluence of 4×10^{23} n/cm^2. Third, no blanket zone may surpass 3 at.% burnup, as per the EBR-II Mark-II fuel pin tests. Fourth, no driver fuel may exceed 13.33 at.% burnup; this is reduced from the 20 at.% reached in the IFR pin tests because whereas most fast reactor fuel pins have a plenum-to-fuel volume ratio of unity, the SABrR pins have a ratio of only 2/3.

There were also soft constraints placed on each case, which were considered more as design guidelines. If a soft constraint is violated, the scenario may be continued either if the violation is temporary or if a scenario is approaching the violation of a hard constraint. There were two soft constraints. First, k_{eff} should be significantly below 1 ($k_{eff} < 0.95$ was desired), such that $r \gg \beta$, and the reactor is always very far from prompt critical. Second, the output of the fission core plus blankets should be maintained at 3,000 MW(thermal) using a maximum of 500 MW of fusion power, not much greater than the ITER design DT fusion power level.

24.2 Results and discussion

24.2.1 TBR case

The configuration of reflector and tritium breeding blanket that emphasizes a high TBR is shown in relation to the fission core and plasma in Figure 24.2. (Other structures that are not important in determining the TBR are omitted for clarity.) Placing the outboard tritium breeding blanket adjacent to the fission annulus results in a higher neutron capture rate in the blanket than if it were located radially outside the reflector. However, the increase in neutron capture comes at the expense of some of the fissile breeding in the outer radial fissile blanket.

The k_{eff}, k_{mult}, and fusion power required to drive the fission annulus at 3,000 MW(thermal) are shown in Figure 24.3. Shortly before reaching 2,000 days

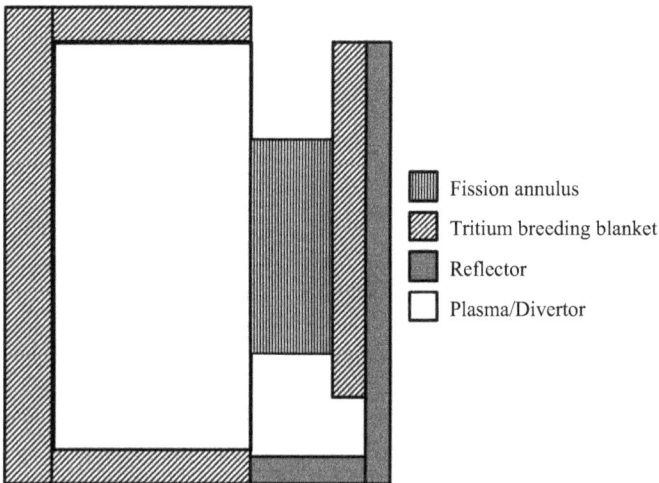

Figure 24.2 *Configuration showing TBR factors (other reactor structures omitted)*

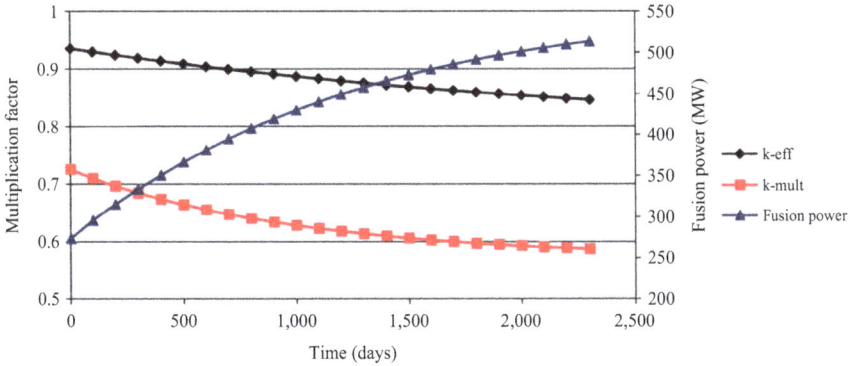

Figure 24.3   Multiplication values and fusion power for TBR case

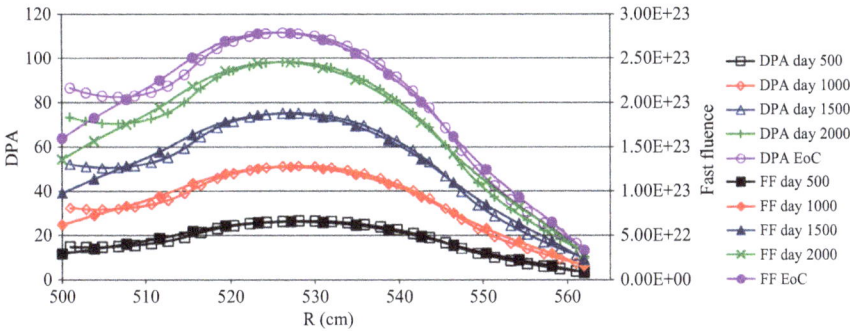

Figure 24.4   Accumulated DPA and fast fluence across core midplane in TBR configuration

of fuel residence time, the fusion power exceeds 500 MW, but the blanket fuel in the plasma-side edge of the inner radial blanket reaches 3% burnup soon after day 2300, at a fusion power of 513 MW. The maximum burnup in the driver fuel is 9.31%, well below its burnup limit of 13.33%. The TBR is substantially above 1.15 for the entire cycle, and the FBR is 1.299 at its peak and 1.278 at the end of the fuel residence time. The average net fissile production over the residence time is 208.4 kg/year.

The fast fluence ($E_n > 0.1$ MeV) and DPA accumulation in the cladding across the fission core midplane are shown in Figure 24.4 at various points throughout the fuel life; the end of cycle (EoC) is after 2,300 days. The contribution of the 14.1-MeV fusion neutrons to the total radiation damage can be seen in the upturn of the DPA curve near $R = 500$ cm.

Because unmoderated fusion neutrons are far more damaging than the average fission neutron in the core, while the fast fluence tallies all neutrons above 0.1 MeV equally, the two curves diverge at the plasma source, despite their agreement throughout the rest of the core. This difference is more pronounced later in the core residence time when the source power has been turned up, but the peak for both measures of radiation damage still lies near the midpoint of the fission core and is well below the design limits.

24.2.2   FBR case

The configuration of reflector and tritium breeding blanket that emphasizes a higher fission breeding ratio (FBR) is shown in relation to the plasma and fission core in Figure 24.5.

The k_{eff}, k_{mult}, and fusion power required to drive 3,000 MW(thermal) in the FBR case are shown in Figure 24.6. The higher multiplication values and the lower fusion power are due to fewer net neutrons leaking radially outward from the outer radial blanket zone than in the case favoring the TBR. This directly causes both more power and more fissile production in that assembly ring and indirectly increases those values in the adjacent driver fuel. The limiting factor for fuel residence time in this configuration is, as in the TBR case, the burnup limit of the plasma-edge blanket fuel being reached. However, because of the relatively lower fusion power throughout the entire residence time and the consequently lower contribution to the 3,000-MW(thermal) fission output from that zone, it took 2,600 days to reach the limit. The driver fuel has a maximum burnup of 10.23%, comfortably below its maximum. The TBR, while lower in this case than in the case

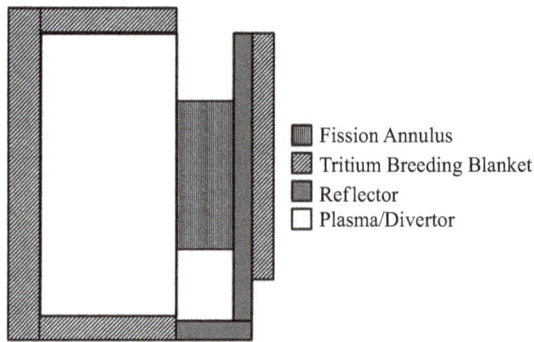

Figure 24.5   Configuration emphasizing FBR (other reactor structures omitted)

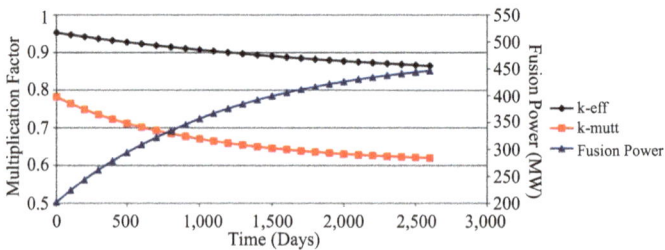

Figure 24.6   Multiplication values and fusion power for FBR case

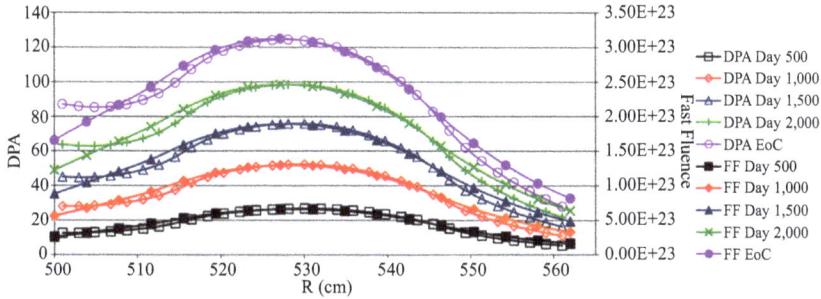

Figure 24.7   Accumulated DPA and fast fluence across core midplane in FBR configuration

favoring tritium production, is 1.206 at its lowest (this occurs at the EoC), with most of the difference resulting from decreased production in the outboard tritium breeding blanket. The FBR is 1.34 at its peak and 1.298 at the EoC. An average net of 253.7 kg/year of fissile material is produced each year in this configuration.

The fast fluence and DPA accumulation across the core midplane are shown in Figure 24.7 at various points throughout the fuel life; the EoC occurs after 2,600 days. Similarly to the TBR case, the fusion neutrons cause a divergence of the DPA and fast fluence near the plasma, but the peaks of both curves are near the annulus center. The maximum DPA is 124, and the maximum fast fluence is 3.12×10^{23} n/cm^2.

24.2.3   Neutronic effect of insulating sheath

A sensitivity study was performed on the FBR configuration to evaluate the effects of removing the LiNbO$_3$ insulating sheath from around each fuel pin. The motivation for doing so stems from the desire to compare with critical fast breeder reactors, which do not require the insulator. While oxide-fueled reactors will have oxygen present in greater fractional quantities than SABrR, Li is absent in even those cores and represents a moderating element unique to SABrR. Though the insulating sheath is less dense than the cladding, it occupies 9.5% of the cross-sectional area within each fuel assembly, so its effect is non-negligible.

For this study, the reactor geometry is otherwise identical to the FBR configuration, and the enrichment of the driver fuel is kept the same. Because the base FBR configuration was able to run until day 2600 before violating one of the design constraints, the sensitivity study was carried out for the same duration, regardless of violation of design constraints. The effects of removing the sheath from around the fuel pins are summarized in Table 24.1.

The removal of the sheath hardened the neutron spectrum. The fission-to-capture ratios of the fuel consequently rose, resulting in a more reactive fission annulus that was easier to drive with the neutron source. The higher fission-to-capture ratios of the fissile isotopes meant a lower destruction rate for a given fission power, which increases the FBR; however, this effect was more than offset by the increase in

Table 24.1   Effects of removing insulating sheath

Quantity	FBR configuration base case	LiNbO$_3$ sheath removed
BoC k_{eff}[a]	0.953	0.971
EoC k_{eff}	0.865	0.879
BoC k_{mult}	0.781	0.870
EoC k_{mult}	0.619	0.666
BoC P_{fus} (MW)	202	108
EoC P_{fus} (MW)	446	364
FBR (peak/EoC)	1.34/1.298	1.301/1.277
Fissile gain (kg/year)	253.7	212.4
TBR (minimum)	1.206	1.366
Peak blanket burnup (at.%)	2.98	2.43
Peak driver burnup (at.%)	10.23	10.48

[a]BoC, beginning of cycle.

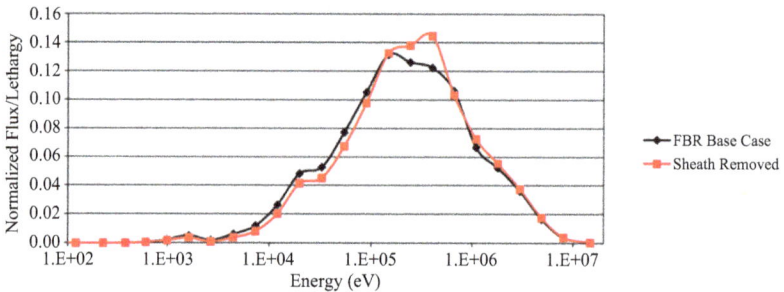

Figure 24.8   Neutron (core-center) spectrum comparison of sheathed and unsheathed pins

leakage from the fission annulus, so the resulting FBR is slightly lower due to reduced capture in fertile isotopes. This increase in leakage from the fission annulus is also evident in the higher TBR. Though the demand on the fusion neutron source, and thus the tritium destruction rate, is lower without the sheath, the majority of the tritium production occurs in the blankets surrounding the plasma, whose production is highly dependent on the neutron source strength (Figures 24.8 and 24.9).

24.2.4   Neutron spectra comparison

The neutron spectra at several points in the core for the TBR and FBR cases are shown in Figures 24.9 and 24.10, respectively. At 1,000 days into the fuel residence time, the spectrum in the inner radial blanket only 2 cm away from the plasma demonstrates the effect of the neutron source on the overall spectrum. The spectra are similar throughout the fission annulus; however, in the outboard tritium breeding blanket, there is a pronounced softening of the spectrum. The presence of the reflector between the fission annulus and the tritium breeding blanket in the FBR case significantly enhances this softening of the spectrum relative to the TBR case.

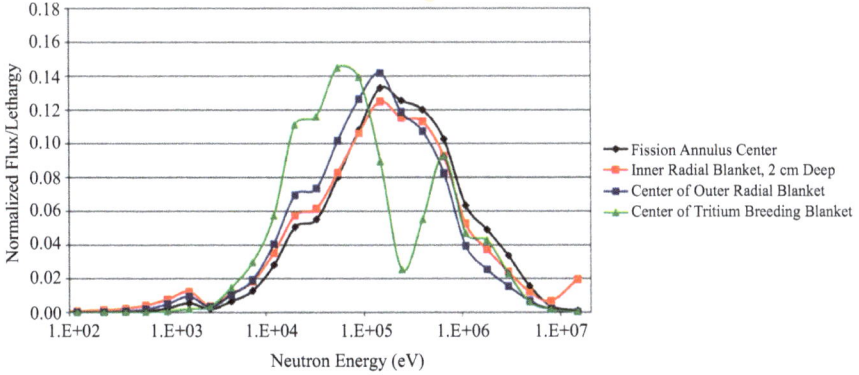

Figure 24.9   Neutron spectra at 1,000 days at selected locations (TBR case)

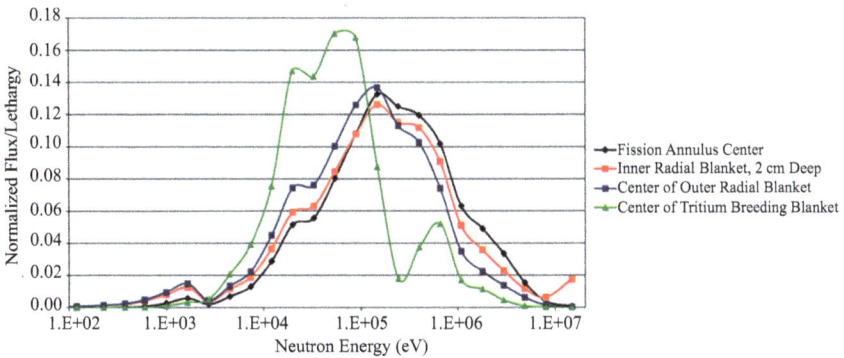

Figure 24.10   Neutron spectra at 1,000 days at selected locations (FBR case)

24.2.5   Power distributions

The distribution of power produced in the driver and in the fission blankets changes significantly as burnup progresses. Initially, the driver fuel produces nearly all of the fission power, but as fissile isotopes are depleted from the driver fuel and bred in the blankets, the blankets produce an increasing fraction of the power. This increase in blanket power is more pronounced in the inner radial blanket assemblies than the outer ones, as they are exposed directly to the fusion neutron source and thus have a higher rate of breeding and a high incident neutron flux from the plasma. The radial power distribution for the FBR case is shown for various times in the burnup cycle in Figure 12.11.

The TBR case power distribution is almost identical to that of the FBR case. The power at any given time in the burnup cycle is slightly higher in the inner radial blanket than in the FBR due to the comparatively higher incident fusion neutron flux. The outer radial blanket power is slightly lower in the TBR case due

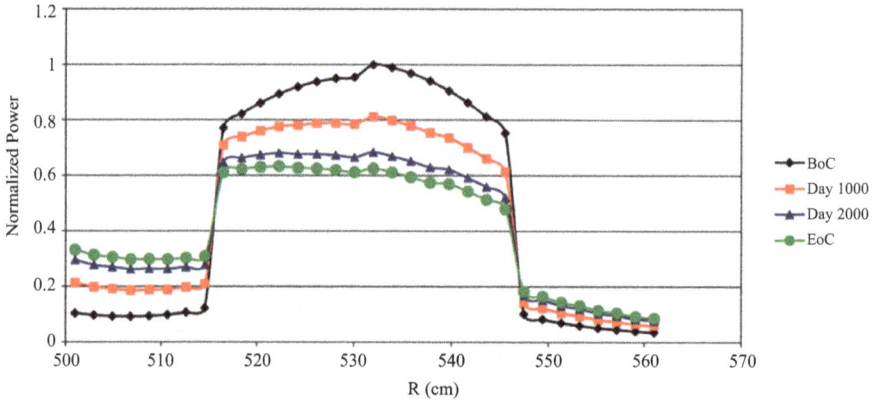

Figure 24.11   Radial power distribution across centerline of the fission annulus at various residence times (FBR case)

to competing neutron capture in the adjacent tritium breeding blanket suppressing fissile production and reducing the neutron flux in that region.

24.2.6   Comparison to critical fast reactor system

A comparison of the breeding performance of SABrR with the high-breeding metal-fueled S-PRISM core design [18] from which the SABrR fuel pins were adapted is shown in Table 24.2. This critical system was chosen for the comparison because of the pin similarity and because of the maturity of the S-PRISM design.

The lower specific power and higher TRU loading of SABrR are a direct consequence of the annular geometry of its fission core; such geometry has a much higher leakage than the traditional pancaked cylinder, so the driver fuel k_∞ must be correspondingly higher, even for a lower k_{eff}.

The higher fissile loading of SABrR means that despite its higher FBR, it has a longer doubling time. However, the fissile gain normalized to fission core thermal power is roughly equal for the SABrR TBR case and S-PRISM, while the SABrR FBR case exceeds S-PRISM in this regard.

SABrR has a higher fuel residence duration than the driver fuel of S-PRISM but a slightly lower blanket residence time. This S-PRISM core design is radially heterogeneous and utilizes blanket shuffling to flatten the radial power profile. SABrR, however, does not shuffle assemblies at any point during the burnup cycle; the presence of the neutron source at the edge of the fission annulus and the ability to adjust its strength largely negate the need to do so for power-flattening purposes. *Therefore, SABrR would be shut down far less frequently for shuffling/refueling purposes, which is an advantage it holds over nearly all critical systems.*

Because cycle lengths for both the FBR and TBR cases for SABrR were limited by blanket burnup in ring 1 with a reasonable margin to peak radiation damage and driver burnup limits, the radial blankets in rings 1 and 4 might be switched

Table 24.2   *Breeding performance comparison of SABrR (FFH) and S-PRISM (CFR)*

Item	SABrR	S-PRISM
Core thermal power (MW)	3,000	1,000
BoC [a] Pu loading (kg)	14,317.0	3,159.9
BoC fissile Pu loading (kg)	9738.2	2,458.8
BoC U loading (kg)	164,763.1	33,052.7
Specific power (W/g Pu)	209.54	316.47
Pu enrich [wt%, Pu/(U + Pu)]	Driver zone 1: 22.36 Driver zone 2: 23.75	21.29

	TBR case	FBR case	Unsheathed	—
Fuel residence time (days)	2,300	2,600	2,600	Driver: 2,070 Blanket: 2,760
Cycle-average breeding ratio	1.28	1.32	1.28	1.22
Fissile gain (kg/year)	208.39	253.73	212.37	69.91
Normalized fissile gain [kg/(MW(thermal)-year]	0.0695	0.0846	0.0708	0.0699

[a]BoC, beginning of cycle.

midcycle to increase total residence time, although at the cost of increased downtime. Finding a suitable electrically insulating material that has less moderating power than the $LiNbO_3$ would also extend the cycle duration of SABrR since the blanket burnup limit was not reached in that scenario. A less moderating insulator would allow for either decreased fissile enrichment of the driver fuel or for radially heterogeneous core layouts, which do not increase driver enrichment to high levels.

We note that no attempt has been made to optimize these initial SABrR designs for fissile production within a particular fuel cycle. *In principle, subcritical operation (a) removes the criticality requirement, which allows the fuel and blanket to remain in the reactor until the clad radiation damage limit is reached, and (b) increases the reactivity margin to prompt critical by an order of magnitude, which removes any safety limitation on Pu content in the reactor.* A future investigation will seek to leverage these two factors to improve the fissile production performance of SABrR for comparison against critical fast burner reactors. Our purpose here was to investigate whether the SABR burner reactor concept (technology, geometry, and major parameters) could be adapted to a breeder reactor that had a reasonable fissile production performance, which seems to be the case.

24.3   Conclusion

The SABR FFH fast burner reactor configuration, based on IFR-PRISM fast reactor physics and technology and on ITER fusion physics and technology, was investigated for a fast FFH breeder reactor application. Representative configurations for breeding fissile material from depleted uranium while simultaneously breeding tritium were considered, subject to realistic constraints on (a) the radiation damage to the cladding (200 DPA or 4×10^{23} n/cm^2 fast fluence), (b) driver fuel burnup (13.33 at.%), (c) blanket fuel burnup (3 at.%), and (d) TBR > 1.15). The representative designs considered were found to be capable of producing FBR > 1.3 and maintaining TBR > 1.2. *This neutron economy is sufficient to produce 250 kg/year of fissile material in a 3,000-MW(thermal) plant while also producing enough tritium for self-sufficiency of the fusion neutron source fuel.*

While this study demonstrates the capability of the SABrR FFH fast breeder concept to breed significantly excess fissile material and to maintain the tritium needs of the plant, it has not addressed the safety advantage of a FFH breeder relative to a similar critical breeder, and this will be discussed in a later section in the book.

Chapter 25

Using fusion neutrons to achieve a breeding fission fuel cycle

25.1 Introduction

Closing the nuclear fuel cycle is an important step in advancing the prospects of nuclear energy in both the near and far terms. The once-through cycle largely employed today uses a very small percentage of the potential energy content of natural uranium ore and produces high-level waste, for which we have yet to implement a long-term solution. A solution to the overall fuel cycle problem would have the dual benefits of extending the uranium resources of earth by a factor of 10–100 over the once-through cycle and of greatly reducing the volume, decay heat, and longevity of repository-bound waste. Various fast reactor technologies and designs have been developed with the intent of closing the front end (breeder reactors), the back end (burner reactors), or both, of the fuel cycle.

Breeder reactors take advantage of the high neutron-per-fission yield of fissile isotopes, particularly ^{239}Pu, in fast neutron spectra to supply extra neutrons beyond those necessary to sustain the fission chain reaction. These excess neutrons can be captured in fertile material such that more fissile material is produced than was consumed. Burner reactors leverage a fast neutron spectrum to transmute, preferably by fission, transuranic (TRU) isotopes that remain in the spent fuel discharged from thermal reactors. These TRUs, which constitute a substantial fraction (tens of percent) of burner reactor fuel, would otherwise be sent directly to a geological repository and dominate the long-term radiotoxicity and decay heat of the used nuclear fuel. Some reactor designs incorporate the aspects of both breeders and burners. They are intended to operate in an integrated fuel cycle, mixing their own discharged fuel with used fuel from other reactors and depleted uranium to form the next fuel loading. One of the most mature of these integrated reactor concepts is the integral fast reactor (IFR); many of its design decisions and material choices reflect its integrated fuel cycle and the very hard neutron spectrum which that cycle requires.

A subcritical advanced burner reactor (SABR) concept that addresses the waste problem is being developed at the Georgia Institute of Technology. SABR is a sodium-cooled, 3,000 MW (thermal) annular fast reactor consisting of four assembly rings surrounding a toroidal plasma. The fission core operates in the subcritical regime; the plasma supplies an external neutron source via the D–T fusion reaction. The fuel pins of SABR are loaded with TRUs processed from used fuel from light water reactors (LWRs) that has been fissioned to a high atomic percent burnup. Neutrons leaking from the fission core are captured in surrounding tritium breeding blankets to produce fuel for the fusion reaction. Whereas the margin of safety against a prompt critical power excursion is related to the delayed neutron fraction of the fissionable fuel isotope in a critical reactor, the subcriticality margin can be much larger in a subcritical reactor, thus enabling a subcritical reactor to be fueled entirely with TRU, rather than a mixture of TRU with U235, which has a larger delayed neutron fraction than TRU. Fuel cycle calculations indicate that SABR could consume TRUs at triple the rate that an LWR of the same power output produces them; a future reactor fleet might then produce 75% of its electricity in LWRs and 25% in SABRs and send no TRUs, other than losses from reprocessing, to repositories. SABR is based on existing nuclear technologies developed for the IFR and on ITER physics and fusion technology, and could be deployed by the midcentury.

Because the plasma and technology performance required for an economical fusion power plant significantly exceeds that which will be demonstrated in ITER, developing fission–fusion hybrid (FFH) reactors with ITER-level plasma and technology requirements in parallel with the further plasma and technology development needed for pure fusion power would allow for substantial accumulation of power reactor-operating experience with tokamaks prior to the introduction of pure fusion power plants into reactor fleets. In the near term, these FFHs would likely be devoted to burning actinides, while in the longer term, using them for fissile production becomes more desirable as easy-to-extract ^{235}U becomes depleted. The SABR studies to date indicate the efficacy of the FFH in the burner role. An important question then arises: Could a similar hybrid reactor make a useful contribution as a breeder?

Moir *et al.* [10,13] have explored a different (mirror) FFH breeder in some depth, investigating economic scenarios and materials choices with respect to different fertile isotopes and LWR support ratios (see Section 17.5). Those studies focused on a tandem mirror fusion device with substantially different geometry than SABR. Nevertheless, the important considerations are the same as for a tokamak-geometry fast breeder FFH. A key challenge in designing an effective FFH breeder is the neutron economy: Of the neutrons released from each fission, somewhat less than one must go toward sustaining the chain reaction, a fraction are captured parasitically in fissile material and structural materials, a fraction leak out of the reactor, and the remainder are available for absorption in fertile material to breed fissile material and fuel for the fusion reaction.

The purpose of this section is to investigate if the basic SABR configuration and fusion physics and technology can be effectively used to breed fissile material. Instead of the TRU fuel used in the SABR burner reactor, the fuel pins for the subcritical advanced breeder reactor (SABrR) contain U-Pu-Zr and U-Zr metal fuel. The primary difference in neutronics design challenges between the burner and breeder SABRs revolves around the neutron economy. Whereas in the burner reactor only the total fission rate and the tritium breeding ratio (TBR) are important, in the breeder reactor, both the TBR and the fissile breeding ratio (FBR) are important. Previous calculations indicate that a TBR of at least 1.15 must be maintained to provide for tritium self-sufficiency of the fusion neutron source, and a FBR of significantly greater than unity must be achieved to provide fissile material for other reactors. The challenge of keeping a sufficiently high TBR in SABrR is exacerbated by the presence of U-Zr fissile breeding blankets between the annular fission core and the surrounding tritium breeding blanket, which significantly reduces the neutron flux incident on the tritium breeding blankets in the breeder relative to the burner. With this in mind, two configurations of the tritium blanket and reflector structures were considered: one that maximizes the TBR and one that maximizes the FBR.

Ultimately, any FFH design must compare favorably to critical fast fission reactor designs to justify the added cost and complexity of including the fusion element. Such a comparison must include comparative analyses of (a) the fissile production capability of subcritical and critical fast breeder reactors, (b) the dynamic responses of the subcritical and critical designs to various accident scenarios, (c) the overall cost and reliability of the system of breeder reactors and reprocessing/refabrication facilities, and (d) the resistance to proliferation. The first step in this evaluation is to identify a realistic subcritical (FFH) fast breeder reactor technology for comparison with critical fast breeder reactor concepts.

This section establishes the core design of two variants of the SABrR geometry and evaluates their capability to breed fissile material within several constraints established for the SABR design; future studies on the safety performance of SABrR will be based on this design. Although no effort has been made to optimize the fissile breeding performance of these initial SABrR designs within a particular fuel cycle, a comparison of breeding performance and fuel cycles is made with a critical high-breeding, metal-fueled S-PRISM core design to illustrate the feasibility of SABrR.

25.2 SABrR design concept

The top-level configuration of the SABR burner concept is shown in Figure 25.1, and a detailed R-Z cross section is shown in Figure 25.2. The entirety of the fusion and fission systems resides within the superconducting toroidal magnets of the tokamak. The inner edge of the annular fission core lies at the outer edge of the tokamak plasma chamber wall. Surrounding the plasma chamber and fission core annulus are first the tritium breeding blankets and then the stainless steel reflector. Finally, these are enveloped by multilayer shields that reduce the fast neutron and gamma fluences to the superconducting magnets, giving them a lifetime of at least 30 full power years.

25.2.1 Fusion neutron source

The fusion neutron source is provided by a tokamak based on ITER physics and is capable of 500 MW of fusion power. Deuterium and tritium in the plasma undergo the reaction $^2_1D + ^3_1T \rightarrow ^4_2He + ^1_0n$. The helium nucleus will deposit most of its energy, ~3.5 MeV, into the plasma. However, the neutron, carrying 14.1 MeV, will stream directly out of the tenuous plasma, since it has no charge and is therefore not

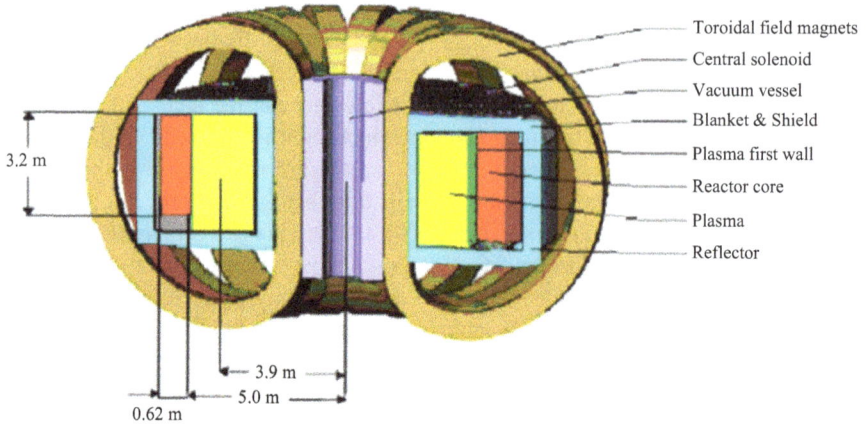

Figure 25.1   SABrR configuration

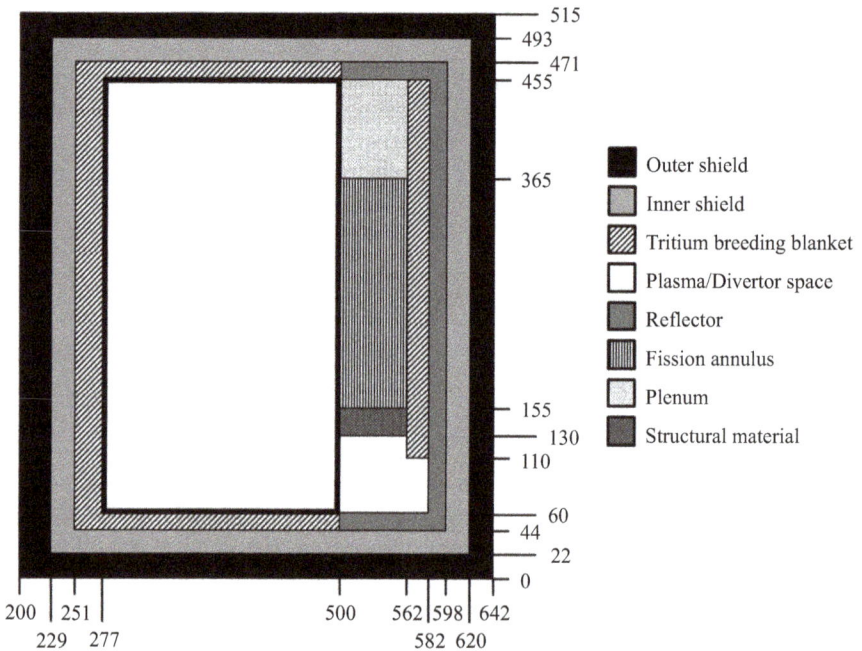

Figure 25.2   R-Z cross section of SABrR (dimensions in units of centimeters)

bound by the magnetic fields confining the plasma. This neutron, possessing energy several times that of the average fission neutron, is extremely well-suited to sustaining a subcritical fission reaction: Not only is the fission-to-capture ratio for heavy metal nuclides higher at such high energies, but the neutron also has energy well in excess of the threshold fission reactions in the even-neutron isotopes, of which ^{238}U, ^{240}Pu, and ^{242}Pu are of primary importance. Furthermore, $(n,2n)$ and $(n,3n)$ reactions contribute substantially to neutron production via the fusion source due to the increase of these cross sections at high neutron energy.

It is not difficult to obtain half of the fuel for the D–T fusion reaction. Deuterium, present in about 1 of every 10,000 water molecules, is relatively easy to recover. Tritium, however, has a half-life of 12.32 years and must therefore be produced. To do so, tritium breeding blankets composed of lithium orthosilicate (Li$_4$SiO$_4$) are placed around the plasma chamber and fission annulus. Lithium occurs naturally in two isotopes: 7% is ^{6}Li and 93% is ^{7}Li. Tritium is therefore produced by the reactions

$$^6_3\text{Li} + ^0_1\text{n} \rightarrow ^3_1\text{T} + ^4_2\text{He}$$

and

$$^7_3\text{Li} + ^1_0\text{n} \rightarrow ^3_1\text{T} + ^4_2\text{He} + ^1_0\text{n}$$

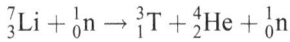

The second reaction is endothermic with a threshold energy of $E_n = 2.466$ MeV, whereas the first reaction is exothermic. This, combined with the high absorption cross section of ^{6}Li for neutrons at thermal energies, causes the ^{6}Li reaction to be far more effective at tritium production despite its much lower isotopic content. The lithium in the SABrR tritium breeding blankets is enriched to 93% ^{6}Li by weight to increase the tritium production rate.

25.2.2   *Annular fast reactor*

Each SABrR fuel assembly is a hexagonal duct measuring 15.5 cm across flats (Figure 25.3). The duct is filled with 10 rings of fuel pins on a hexagonal lattice, for a total of 271 pins per assembly. The pins are separated by a wire wrap at a pitch of

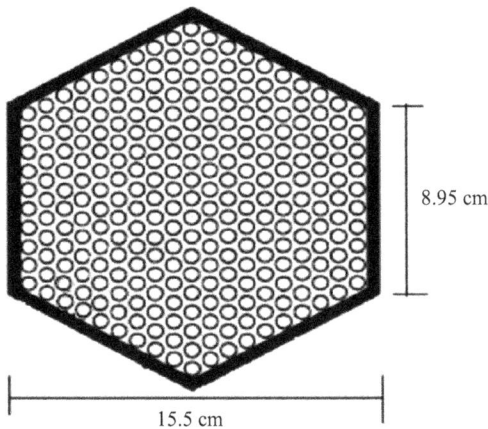

8.95 cm

15.5 cm

Figure 25.3   SABrR fuel assembly

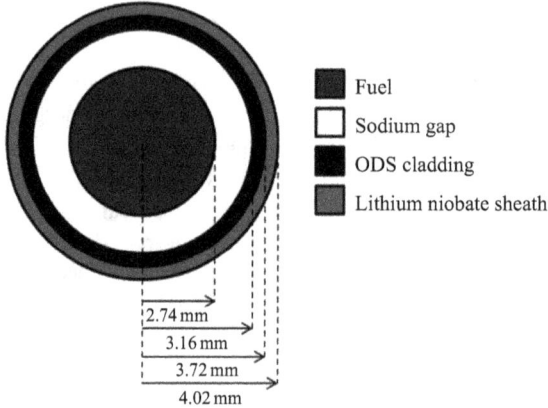

Fuel
Sodium gap
ODS cladding
Lithium niobate sheath

2.74 mm
3.16 mm
3.72 mm
4.02 mm

Figure 25.4   SABrR fuel pin

8.9 mm. The fuel pins are based on the pins developed in the IFR initiative for S-PRISM (Figure 25.4). The cladding is ODS MA957, a ferritic oxide dispersion strengthened (ODS) steel that is estimated to be able to withstand up to 200 displacements per atom (DPA). ODS MA957 was developed as a low-swelling ferritic steel for fast reactor cladding; at low-temperature irradiation ($T < 355°C$), the ductile-to-brittle transformation temperature shifts upward significantly, causing embrittlement as a failure mode at relatively low accumulated radiation damage. For this reason, the lowest cladding temperature in SABrR is ~380°C and is located at the lower edge of the lower axial blanket, where the fast neutron fluence is significantly below average; similarly, the region of maximum fast fluence (at the core midplane) has cladding temperatures well in excess of 400°C. Around the cladding is a $LiNbO_3$ sheath that provides electrical insulation to prevent a large magneto hydrodynamic pressure drop in the liquid metal coolant.

The fission core fuel height is 200 cm, but this is axially expanded to 210 cm in the computational model to account for thermal and irradiation-induced axial swelling of the fuel column, which occurs at very low burnup. The densities are correspondingly adjusted downward to keep the fuel mass constant. An R-Z cross section of the fuel zones in the fission annulus is shown in Figure 25.5. The driver fuel is located in the axially centered 150 cm of the second and third assembly rings. There are 30-cm axial blankets both above and below the driver fuel. Radial blanket assemblies occupy rings 1 and 4.

The driver fuel is a U-Pu-10Zr ternary alloy very similar to several of the pin compositions tested in EBR-II and the fast flux test facility. The fuel slug has a smear density of 75% to allow for burnup-related swelling due to the production of gaseous fission products. These gases cause the fuel to become porous and swell radially until it contacts the cladding. At ~1% burnup, the porosity is high enough that the pores become inter-connected and the fission gases are released to the plenum; further gaseous fission products that are produced do not contribute to further swelling, and thus, high burnups are achievable [22,23]. In some of the test

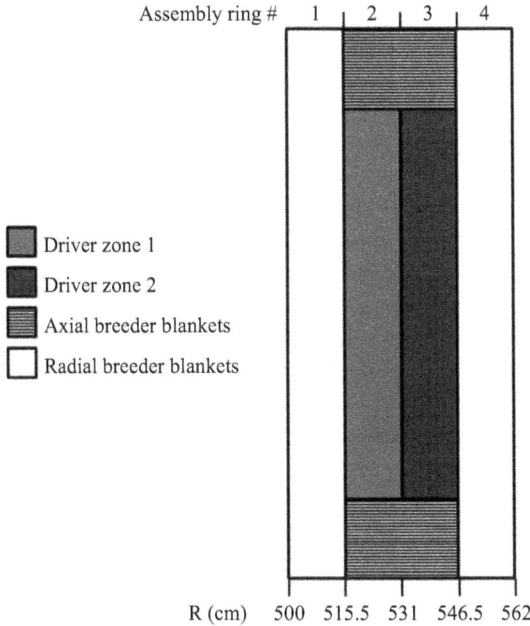

Figure 25.5   SABrR fuel loading

Table 25.1   Plutonium vector of driver fuel

Isotope	Weight percent
^{238}Pu	2.102
^{239}Pu	58.258
^{240}Pu	25.976
^{241}Pu	9.76
^{242}Pu	3.904

pins in EBR-II, burnups of almost 20 at.% were demonstrated in several of the test pins without issue; these tests were ongoing when the reactor was shut down, so 20 at.% can be considered a lower limit of the burnup potential of that fuel. The plutonium vector of the driver fuel is given in Table 25.1; it was developed for high-burnup metal fuels. The plutonium enrichment in driver zone 2 (in the third assembly ring) is 23.75%, slightly higher than in driver zone 1, at 22.36%, to flatten the radial power profile, and it receives fewer fusion neutrons as it is farther away.

The radial blanket fuel is a U-10Zr alloy at 85% smear density; it is similar to the Mark-I fuel pins for EBR-II, except with 10% Zr by weight instead of 5% fissium [23]. The ^{235}U enrichment is 0.25% by weight. The 10% Zr was chosen because of its beneficial effect on the fuel melting temperature and on fuel-cladding interactions. While the lower uranium volume fraction likely has a slightly negative effect on breeding, the plasma-side edge of the inner radial blanket receives the fusion neutrons most directly and thus has a relatively high power for a fast reactor blanket; therefore, the thermal considerations are of primary importance. For that same reason, the pin diameter and number of pins per assembly are kept the same as in the driver fuel rather than using fewer, larger pins found in most blanket assembly designs. The Mark-II fuel pins were limited to 3 at.% burnup in EBR-II due to burnup-induced swelling; these pins are similarly limited. The axial blankets are the same composition as the radial blankets, but at 75% smear density to allow the fission gases from the driver fuel easy access to the plenum. Because of their lower smear density, the burnup limit of the axial blankets matches that of the driver fuel. The breeding of fissile material occurs by neutron capture in fertile isotopes within the fission annulus. The primary fertile isotope is ^{238}U, which captures a neutron and then decays by beta emission twice before becoming fissile ^{239}Pu:

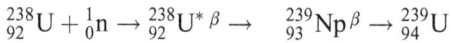

$$^{238}_{92}\text{U} + ^{1}_{0}\text{n} \rightarrow ^{238}_{92}\text{U}^{*} \,^{\beta} \rightarrow \quad ^{239}_{93}\text{Np}^{\beta} \rightarrow ^{239}_{94}\text{U}$$

The decay to ^{239}Np occurs with a half-life of 23 min, and the decay to ^{239}Pu has a half-life of 2.4 days; thus, ^{239}Np reaches near-steady-state levels very shortly after the beginning of the fuel residence. There are two fertile isotopes of Pu as well: ^{238}Pu and ^{240}Pu. Each of these can capture a neutron to become ^{239}Pu and ^{241}Pu, respectively:

$$^{238}_{94}\text{Pu} + ^{1}_{0}\text{n} \rightarrow ^{239}_{94}\text{Pu}$$

and

$$^{240}_{94}\text{Pu} + ^{1}_{0}\text{n} \rightarrow ^{241}_{94}\text{Pu}.$$

Because ^{238}U is the most common isotope of the three in the fresh driver fuel, most of the fissile production is through the first reaction, though all three substantially contribute. In the blanket assemblies, only ^{238}U is initially present, so nearly all fissile production there occurs via ^{239}Np.

25.3   Computational model

Calculation methodology
SABrR was modeled in European Reactor ANalysis Optimized calculation System (ERANOS), a fast reactor code system developed to model the Phénix and SuperPhénix reactors. ERANOS employs the European Cell COde (ECCO) to collapse 1968-group JEFF2.0 cross sections within each reactor lattice cell to the 33 groups used in core calculations, ranging from 20 MeV down to 0.1 eV. The core geometry was described in R-Z cylindrical geometry and the core calculations performed in the ERANOS discrete ordinates transport module BISTRO using an S8 quadrature with 132 radial and 216 axial mesh points. The fuel was depleted for 100 days in each burnup step in the EVOLUTION module before reperforming the core neutron flux calculations.

At each depletion step, the neutron source multiplication k_{mult} is calculated, and the neutron source strength is adjusted such that the fission annulus output is 3,000 MW(thermal). The fusion power P_{fus} required to maintain a given fission power P_{fis} is determined by k_{mult}, the average number of neutrons released per fission n, the energy released per fusion E_{fus}, and the energy released per fission E_{fis}:

$$P_{fus} = P_{fis} \left(\frac{1 - k_{mult}}{k_{mult}} \right) \tag{25.1}$$

It is important to note that k_{mult} differs from the more familiar k_{eff}. The TBR is also calculated at each step to determine if enough tritium is being produced to fuel sustained operation of the fusion neutron source.

The TBR is defined as

$$\mathrm{TBR}(t) = \frac{\int_V \Sigma_c^{Li} \phi(r, t) \mathrm{d}V}{\int_V S(r, t) \mathrm{d}V}, \tag{25.2}$$

where S is the fusion neutron source. This only accounts for the production of T by ^{6}Li capture and thus is a conservative estimate of the TBR, as T produced in the threshold reaction in ^{7}Li is not counted in the ERANOS calculation. However, since the tritium breeding material is highly enriched in ^{6}Li and the cross section for production via that route is much higher, the approximation should be quite close to the true tritium production rate.

The FBR is the instantaneous ratio of the production rate of fissile atoms to their destruction rate, whether through fission or parasitic capture:

$$\mathrm{FBR}(t) = \frac{P(t)}{D(t)}$$

The production rate is calculated by integrating the capture rates of the fertile isotopes over the reactor volume:

$$P(t) = \int_V \left(\Sigma_c^{238U}(r)\phi(r, t) + \Sigma_c^{238Pu}(r)\phi(r, t) + \Sigma_c^{240U}(r)\phi(r, t) \right) \mathrm{d}V. \tag{25.3}$$

Though ^{239}Np, rather than ^{238}U, is technically the precursor to ^{239}Pu, ^{239}Np exists in the reactor in a near steady state after its first few half-lives. Thus, by approximately day 20 of fuel residence time, the decay rate of ^{239}Np is equal to the capture rate of ^{238}U. The destruction rate is the volume-integrated absorption rate for all of the fissile isotopes. Only ^{235}U, ^{239}Pu, and ^{241}Pu exist in substantial amounts in the reactor, so other fissile isotopes are omitted from the summation:

$$D(t) = \int_V \left(\Sigma_c^{235U}(r)\phi(r, t) + \Sigma_{abs}^{239Pu}(r)\phi(r, t) + \Sigma_{abs}^{241Pu}(r)\phi(r, t) \right) \mathrm{d}V. \tag{25.4}$$

Substituting these expressions for the production and destruction rates into (17.5), we have

$$\text{FBR}(t) = \frac{P(t)}{D(t)} = \frac{\int_V \left(\Sigma_c^{238\text{U}}(r)\phi(r,t) + \Sigma_c^{238\text{Pu}}(r)\phi(r,t) + \Sigma_c^{240\text{U}}(r)\phi(r,t) \right) dV}{\int_V \left(\Sigma_c^{235\text{U}}(r)\phi(r,t) + \Sigma_{\text{abs}}^{239\text{Pu}}(r)\phi(r,t) + \Sigma_{\text{abs}}^{241\text{Pu}}(r)\phi(r,t) \right)}$$

(25.5)

25.3.1   Design constraints

There were four hard constraints placed on the reactor design that, if violated, were a termination point for that particular case. First, the TBR must not fall below 1.15. This "practical" value was chosen because tritium self-sufficiency is a requirement for sustained fusion operation and previous calculations indicate that this excess above unity allows for losses due to inefficiency in tritium collection and for the radioactive decay of any tritium in inventory throughout the operating and refueling cycles. Second, the radiation damage limit of the clad must not be exceeded. The damage limit of ODS MA957 in a fission spectrum is estimated at either 200 DPA or at an accumulated fast fluence of 4×10^{23} n/cm^2. Third, no blanket zone may surpass 3 at.% burnup, as per the EBR-II Mark-II fuel pin tests. Fourth, no driver fuel may exceed 13.33 at.% burnup; this is reduced from the 20 at.% reached in the IFR pin tests because whereas most fast reactor fuel pins have a plenum-to-fuel volume ratio of unity, the SABrR pins have a ratio of only 2/3.

There were also soft constraints placed on each case, which were considered more as design guidelines. If a soft constraint is violated, the scenario may be continued either if the violation is temporary or if a scenario is approaching the violation of a hard constraint. There were two soft constraints. First, k_{eff} should be significantly below 1 ($k_{\text{eff}} < 0.95$ was desired), such that $r \gg \beta$, and the reactor is always very far from prompt critical. Second, the output of the fission core plus blankets should be maintained at 3,000 MW(thermal) using a maximum of 500 MW of fusion power, only 25% more than the ITER design DT fusion power level.

25.4   Results and discussion

The configuration of reflector and tritium breeding blanket that emphasizes a high TBR is shown in relation to the fission core and plasma in Figure 25.6. Placing the outboard tritium breeding blanket adjacent to the fission annulus results in a higher neutron capture rate in the blanket than if it were located radially outside the reflector. However, the increase in neutron capture comes at the expense of some of the fissile breeding in the outer radial fissile blanket.

The k_{eff}, k_{mult}, and fusion power required to drive the TBR at 3,000 MW (thermal) are shown in Figure 25.7. Shortly before reaching 2,000 days of fuel residence time, the fusion power exceeds 500 MW, but the blanket fuel in the

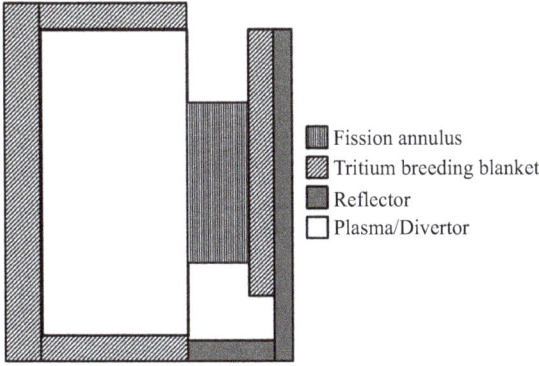

Figure 25.6 Configuration favoring TBR (other reactor structures omitted)

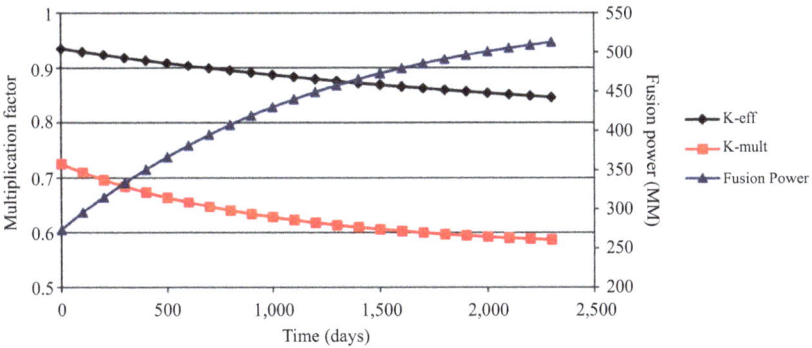

Figure 25.7 Multiplication values and fusion power for TBR case

plasma-side edge of the inner radial blanket reaches 3% burnup soon after day 2300, at a fusion power of 513 MW. The maximum burnup in the driver fuel is 9.31%, well below its burnup limit of 13.33%. The TBR is substantially above 1.15 for the entire cycle, and the FBR is 1.299 at its peak and 1.278 at the end of the fuel residence time. The average net fissile production over the residence time is 208.4 kg/year.

The fast fluence ($E_n > 0.1$ MeV) and DPA accumulation in the cladding across the fission core midplane are shown in Figure 25.8 at various points throughout the fuel life; the end of cycle (EoC) is after 2,300 days. The contribution of the 14.1-MeV fusion neutrons to the total radiation damage can be seen in the upturn of the DPA curve near $R = 5$ (500 cm). Because these unmoderated fusion neutrons are far more damaging than the average fission neutron in the core, while the fast fluence tallies all neutrons above 0.1 MeV equally, the two curves diverge at the plasma source despite their agreement throughout the rest of the core. This differ-ence is more pronounced later in the core residence time when the source power has

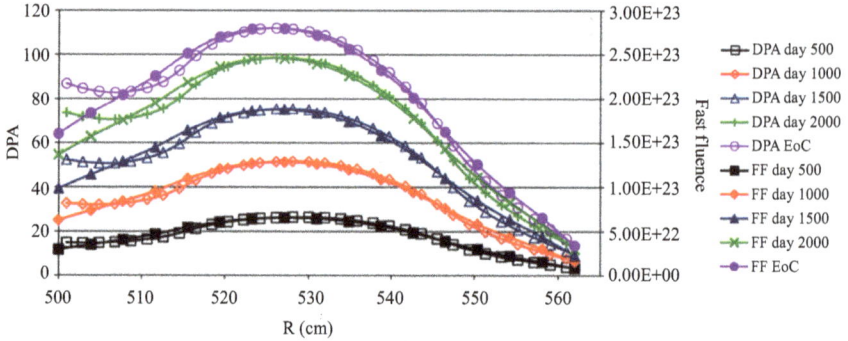

Figure 25.8   Accumulated DPA and fast fluence across core midplane in TBR configuration

been turned up, but the peak for both measures of radiation damage still lies near the midpoint of the fission core and is well below the design limits.

25.4.1   FBR case

The configuration of reflector and tritium breeding blanket that emphasizes a higher FBR is shown in relation to the plasma and fission core in Figure 25.9.

The k_{eff}, k_{mult}, and fusion power required to drive 3,000 MW(thermal) in the FBR case are shown in Figure 25.10. The higher multiplication values and the lower fission power are due to fewer net neutrons leaking radially outward from the outer radial blanket zone than in the case favoring the TBR. This directly causes both more power and more fissile production in that assembly ring and indirectly

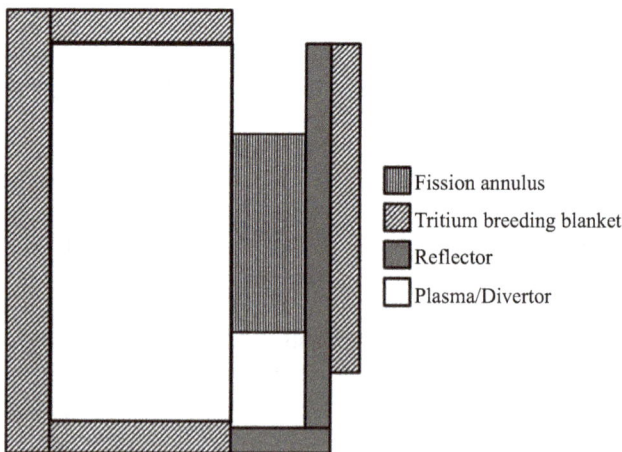

Figure 25.9   Configuration emphasizing FBR (other reactor structures omitted)

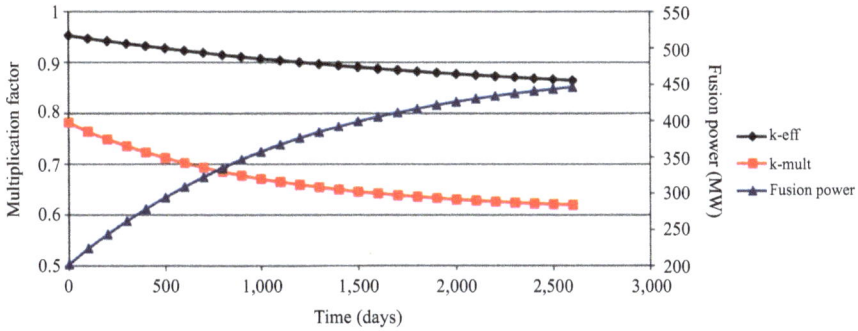

Figure 25.10 Multiplication values and fusion power for FBR case

increases those values in the adjacent driver fuel. The limiting factor for fuel residence time in this configuration is, as in the TBR case, the burnup limit of the plasma-edge blanket fuel being reached. However, because of the relatively lower fusion power throughout the entire residence time and the consequently lower contribution to the 3,000-MW(thermal) fission output from that zone, it took 2,600 days to reach the limit. The driver fuel has a maximum burnup of 10.23%, comfortably below its maximum. The TBR, while lower in this case than in the case favoring tritium production, is 1.206 at its lowest (this occurs at the EoC), with most of the difference resulting from decreased production in the outboard tritium breeding blanket. The FBR is 1.34 at its peak and 1.298 at the EoC. An average net of 253.7 kg/year of fissile material is produced each year in this configuration.

The fast fluence and DPA accumulation across the core midplane are shown in Figure 25.11 at various points throughout the fuel life; the EoC occurs after 2,600 days. Similarly, to the TBR case, the fusion neutrons cause a divergence of the DPA and fast fluence near the plasma, but the peaks of both curves are near the annulus center. The maximum DPA is 124, and the maximum fast fluence is 3.12×10^{23} n/cm^2.

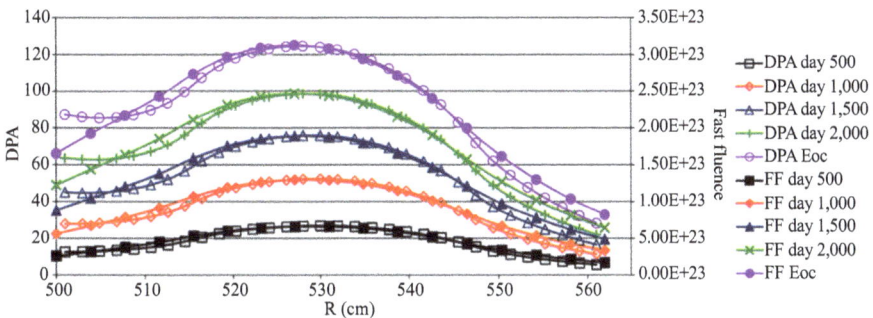

Figure 25.11 Accumulated DPA and fast fluence across core midplane in FBR configuration

25.5   Neutronic effect of insulating sheath

A sensitivity study was performed on the FBR configuration to evaluate the effects of removing the $LiNbO_3$ insulating sheath from around each fuel pin. The motivation for doing so stems from the desire to compare with critical fast breeder reactors, which do not require the insulator. While oxide-fueled reactors will have oxygen present in greater fractional quantities than SABrR, Li is absent in even those cores and represents a moderating element unique to SABrR. Though the insulating sheath is less dense than the cladding, it occupies 9.5% of the cross-sectional area within each fuel assembly, so its effect is non-negligible.

For this study, the reactor geometry is otherwise identical to the FBR configuration, and the enrichment of the driver fuel is kept the same. Because the base FBR configuration was able to run until day 2600 before violating one of the design constraints, the sensitivity study was carried out for the same duration, regardless of violation of design constraints. The effects of removing the sheath from around the fuel pins are summarized in Table 25.2.

The removal of the sheath hardened the neutron spectrum (Figure 25.12). The fission-to-capture ratios of the fuel consequently rose, resulting in a more reactive fission annulus that was easier to drive with the neutron source. The higher fission-to-capture ratios of the fissile isotopes meant a lower destruction rate for a given fission power, which increases the FBR; however, this effect was outcompeted by the increase in leakage from the fission annulus, so the resulting FBR is slightly lower due to reduced capture in fertile isotopes. This increase in leakage from the fission annulus is also evident in the higher TBR. Though the demand on the fusion neutron source, and thus the tritium destruction rate, is lower without the sheath, the majority of the tritium production occurs in the

Table 25.2   Effects of removing insulating sheath

Quantity	FBR configuration base case	$LiNbO_3$ sheath removed
BoC k_{eff} [a]	0.953	0.971
EoC k_{eff}	0.865	0.879
BoC k_{mult}	0.781	0.870
EoC k_{mult}	0.619	0.666
BoC P_{fus} (MW)	202	108
EoC P_{fus} (MW)	446	364
FBR (peak/EoC)	1.34/1.298	1.301/1.277
Fissile gain (kg/year)	253.7	212.4
TBR (minimum)	1.206	1.366
Peak blanket burnup (at.%)	2.98	2.43
Peak driver burnup (at.%)	10.23	10.48

[a]BoC, beginning of cycle.

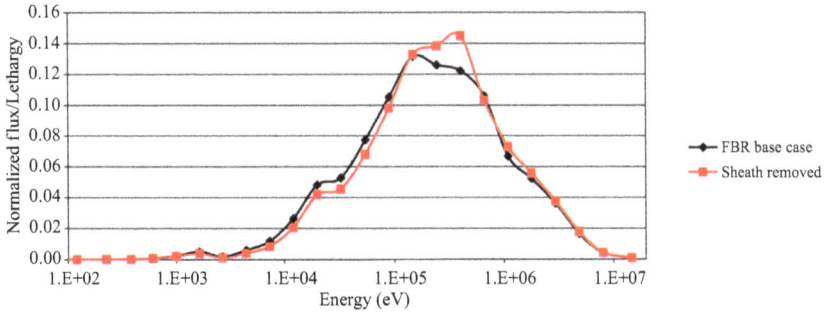

Figure 25.12   Neutron (core-center) spectrum comparison of sheathed and unsheathed pins

blankets surrounding the plasma, whose production is highly dependent on the neutron source strength.

25.5.1   Neutron spectra comparison

The neutron spectra at several points in the core for the TBR and FBR cases are shown in Figures 25.13 and 25.14, respectively, at 1,000 days into the fuel residence time; the spectrum in the inner radial blanket only 2 cm away from the plasma demonstrates the effect of the neutron source on the overall spectrum. The spectra are similar throughout the fission annulus; however, in the outboard tritium breeding blanket, there is a pronounced softening of the spectrum. The presence of the reflector between the fission annulus and the tritium breeding blanket in the FBR case significantly enhances this softening of the spectrum relative to the TBR case.

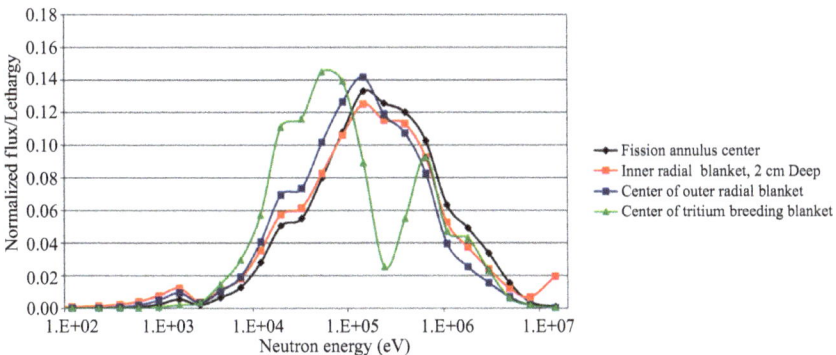

Figure 25.13   Neutron spectra at 1,000 days at selected locations (TBR case)

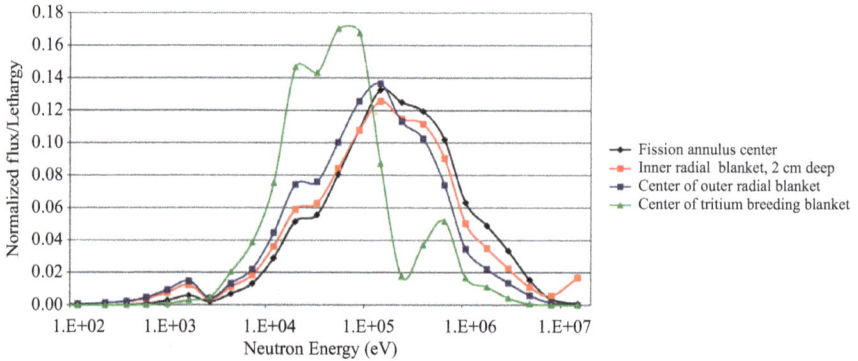

Figure 25.14   Neutron spectra at 1,000 days at selected locations (FBR case)

25.5.2   Power distributions

The distribution of power produced in the driver and in the fission blankets changes significantly as burnup progresses. Initially, the driver fuel produces nearly all of the fission power, but as fissile isotopes are depleted from the driver fuel and bred in the blankets, the blankets produce an increasing fraction of the power. This increase in blanket power is more pronounced in the inner radial blanket assemblies than the outer ones, as they are exposed directly to the fusion neutron source and thus have a higher rate of breeding and a high incident neutron flux from the plasma. The radial power distribution for the FBR case is shown for various times in the burnup cycle in Figure 25.15.

The TBR case power distribution is almost identical to that of the FBR case. The power at any given time in the burnup cycle is slightly higher in the inner radial blanket than in the FBR due to the comparatively higher incident fusion

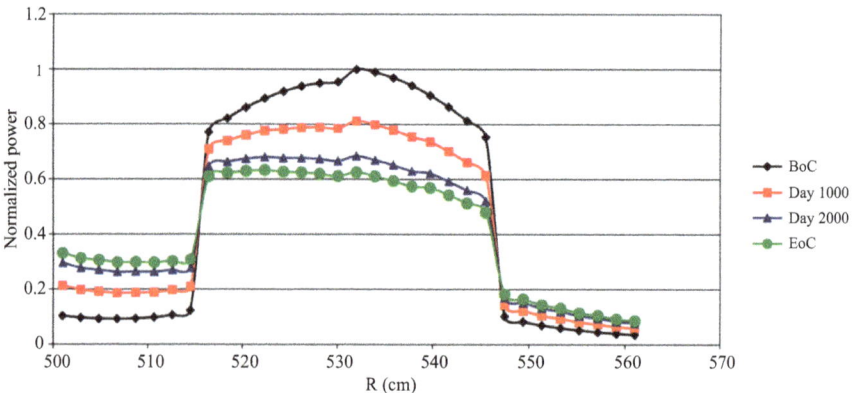

Figure 25.15   Radial power distribution across centerline of the fission annulus at various residence times (FBR case)

neutron flux. The outer radial blanket power is slightly lower in the TBR case due to competing neutron capture in the adjacent tritium breeding blanket suppressing fissile production and reducing the neutron flux in that region.

25.5.3   Comparison with critical fast reactor system

A comparison of the breeding performance of SABrR with the high-breeding metal-fueled S-PRISM core design [18] from which the SABrR fuel pins were adapted is shown in Table 25.3. This critical system was chosen for the comparison because of the pin similarity and because of the maturity of the S-PRISM design.

The lower specific power and higher TRU loading of SABrR are a direct consequence of the annular geometry of its fission core; such geometry has a much higher leakage than the traditional pancaked cylinder, so the driver fuel k_∞ must be correspondingly higher, even for a lower k_{eff}. The higher fissile loading of SABrR means that despite its higher FBR, it has a longer doubling time. However, the fissile gain normalized to fission core thermal power is roughly equal for the SABrR TBR case and S-PRISM, while the SABrR FBR case exceeds S-PRISM in this regard.

SABrR has a higher fuel residence duration than the driver fuel of S-PRISM but a slightly lower blanket residence time. This S-PRISM core design is radially heterogeneous and utilizes blanket shuffling to flatten the radial power profile. SABrR, however, does not shuffle assemblies at any point during the burnup cycle; the presence of the neutron source at the edge of the fission annulus and the ability to adjust its strength largely negate the need to do so for power-flattening purposes. Therefore, *SABrR would be shut down far less frequently for shuffling/refueling purposes, which is an advantage it holds over nearly all critical systems.*

Table 25.3   Breeding performance comparison of SABrR and S-PRISM

Item	SABrR			S-PRISM
Core thermal power (MW)	3,000			1,000
BoC Pu loading (kg)[a]	14,317.0			3,159.9
BoC fissile Pu loading (kg)	9738.2			2,458.8
BoC U loading (kg)	164,763.1			33,052.7
Specific power (W/g Pu)	209.54			316.47
Pu enrich [wt%, Pu/(U + Pu)]	Driver zone 1: 22.36			21.29
	Driver zone 2: 23.75			
	TBR case	**FBR case**	**Unsheathed**	**—**
Fuel residence time (days)	2,300	2,600	2,600	Driver: 2,070
				Blanket: 2,760
Cycle-average breeding ratio	1.28	1.32	1.28	1.22
Fissile gain (kg/year)	208.39	253.73	212.37	69.91
Normalized fissile gain [kg/(MW(thermal)-year)]	0.0695	0.0846	0.0708	0.0699

[a]BoC beginning of cycle.

Because cycle length for both the FBR and TBR cases for SABrR were limited by blanket burnup in ring 1 with a reasonable margin to peak radiation damage and driver burnup limits, the radial blankets in rings 1 and 4 might be switched mid-cycle to increase total residence time, although at the cost of increased downtime. Finding a suitable electrically insulating material that has less moderating power than the $LiNbO_3$ would also extend the cycle duration of SABrR since the blanket burnup limit was not reached in that scenario. A less moderating insulator would allow for either decreased fissile enrichment of the driver fuel or for radially heterogeneous core layouts, which do not increase driver enrichment to high levels.

We note that no attempt has been made to optimize these initial SABrR designs for fissile production within a particular fuel cycle. *In principle, subcritical operation (a) removes the criticality requirement, which allows the fuel and blanket to remain in the reactor until the clad radiation damage limit is reached, and (b) increases the reactivity margin to prompt critical by an order of magnitude, which removes any safety limitation on Pu content in the reactor.* A future investigation will seek to leverage these two factors to improve the fissile production performance of SABrR for comparison against critical fast burner reactors. Our purpose here was to investigate whether the SABR burner reactor concept (technology, geometry, and major parameters) could be adapted to a breeder reactor that had a reasonable fissile production performance, which seems to be the case.

The SABR FFH fast burner reactor configuration, based on IFR-PRISM fast reactor physics and technology and on ITER fusion physics and technology, was investigated for a fast FFH breeder reactor application. Representative configurations for breeding fissile material from depleted uranium while simultaneously breeding tritium were considered, subject to realistic constraints on (a) the radiation damage to the cladding (200 dpa or 4×10^{23} n/cm^2 fast fluence), (b) driver fuel burnup (13.33 at.%), (c) blanket fuel burnup (3 at.%), and (d) TBR > 1.15. The representative designs considered were found to be capable of producing FBR > 1.3 and maintaining TBR > 1.2. *This neutron economy is sufficient to produce 250 kg/year of fissile material in a 3,000-MW(thermal) plant while also producing enough tritium for self-sufficiency of the fusion neutron source fuel.*

While this study demonstrates the capability of the SABrR FFH fast breeder concept to breed a significant excess of fissile material and to maintain the tritium needs of the plant, it has not addressed the safety advantage of a FFH breeder relative to a similar critical breeder, and this will be discussed in the next chapter in the book.

Chapter 26
Dynamic safety analyses of FFH reactors

It is standard practice to analyze the dynamic response of nuclear fission reactors to unanticipated malfunctions. A set of "design basis accidents" have been developed for this purpose: loss-of-coolant accident (LOCA), loss-of (coolant)-flow accident (LOFA), loss-of-power accident (LOPA), etc. We have examined the consequences of such design basis accidents in SABR#1 (Na loop cooling), in SABR#2 (Na pool cooling), and in SABR#3 (10-node calculation model of SABR#2).

26.1   Sodium loop-cooled fast reactor (SABR#1)

Dynamic accident analyses for a Na-loop type SABR#1 are reported in detail in [46] and summarized here. The SABR#1 reactor is a Na-cooled, loop-type, metal-fueled fast transmutation reactor described in Sections 17.1 and 17.3. Outside the toroidal plasma chamber on the outboard side is a 62-cm thick annular rectangular fission core. Both the fusion plasma and fission core are surrounded by reflector, tritium breeding blanket, and shield—a combined radial thickness of 80 cm. The 3.2-m tall fission core is composed of 2 m of active fuel, a 1-m fission gas plenum, and a 20-cm axial reflector capping the fuel pins. There are 271 fuel pins in each of the 918 hexagonal fuel assemblies. In addition to the 902 fuel assemblies, there are 16 control rod assemblies that provide about –9$ of negative reactivity. SABR's fusion neutron source is capable of producing up to 500 MW of thermal energy, and SABR requires 73 MW of auxiliary heating at beginning of life (BOL), 242 MW at beginning of equilibrium cycle (BOC), and 370 MW at end of equilibrium cycle (EOC) to maintain the combined fusion plus fission power output of 3,000 MWt.

26.2   Introduction (SABR#1)

As the United States continues expanding its fleet of nuclear reactors, there will be increasing pressure to deal with the rapidly growing quantity of nuclear waste. One possibility is to recycle the long-lived transuranic (TRU) isotopes in spent nuclear fuel (SNF) and put them back into reactors as fuel. By separating the TRUs—many with decay times far greater than 100,000 years—from the SNF discharged by light water reactors, it is possible to fuel advanced burner reactors while reducing the amount of long-lived SNF that must be stored as high-level waste. TRUs with enormously long decay times could be transmuted into fission products, most of which have half-lives of less than a few hundred years.

Instead of the traditional once-through nuclear fuel cycle currently employed in the United States, repeated reprocessing and recycling of spent fuel and the remaining TRU would significantly decrease the amount of waste that must be stored long after the reactor is shut down. By using a subcritical reactor with a neutron source, longer fuel residence times and hence far deeper burns of the TRU can be achieved because in subcritical reactors the neutron source can be increased to compensate for the loss of neutron multiplication. With a strong-enough neutron source, the material radiation damage limits become the limit for the fuel cycle. No longer constrained by criticality, subcritical reactors can obtain significantly longer fuel residence times.

One such subcritical transmutation reactor concept is the subcritical advanced burner reactor (SABR)#1 that has been developed at the Georgia Institute of Technology [12,26]. SABR#1 is a loop-type sodium-cooled fast reactor with a tokamak deuterium–tritium (DT) fusion neutron source that is capable of burning up to 25% of the TRU over a 7.7-year fuel residence time. The amount of TRU burned in SABR in a single-fuel residence time is limited by the radiation damage accumulation in the structural and cladding materials, but with repeated reprocessing and refabrication, more than 90% of the TRU could be burned.

SABR's variable-strength tokamak D–T fusion neutron source is based on the physics and technology that will be demonstrated in the International Thermonuclear Experimental Reactor [35] (ITER), which began operation in 2017 and will be capable of producing up to 500 MW of power. Since ITER will serve as a prototype for the SABR neutron source, SABR could be operational within 15–20 years after ITER.

In general, fast transmutation reactors fueled entirely with TRU would not have the large negative Doppler coefficient provided by ^{238}U and would have a smaller delayed neutron fraction than a reactor fueled with ^{235}U and ^{238}U, resulting in some possible safety issues. The smaller delayed neutron fraction would lead to a smaller reactivity margin to prompt criticality. However, because subcritical reactors operate at such a large negative reactivity, negative 9.26\$ initially in the case of SABR where $b = 3.009 \times 10^{-3}$, the margin to prompt criticality is increased dramatically, providing subcritical reactors with a major safety advantage, relative to critical reactors, against inadvertent reactivity insertions. The introduction of a neutron source does, however, introduce additional possible safety issues.

We have undertaken a set of dynamic safety analyses of the SABR subcritical fast burner reactor concept in order to investigate the dynamic safety characteristics of subcritical TRU-fueled fast reactors subjected to a variety of possible accident initiation events. The initial studies [45] emphasized the modeling of the fusion neutron source dynamics and the reactor neutron dynamics using a relatively simple representation of the reactor heat removal system. The definition of the SABR core heat removal system including an intermediate heat exchanger (IHX) has now been developed. The coupled dynamics of the reactor core with a fixed neutron source and the sodium heat removal system has been modeled with the RELAP5-3D reactor dynamics code. The purpose of this section is to report the results of this initial study of the accident dynamics of a subcritical, TRU-fueled, sodium-cooled fast reactor with a tokamak D–T fusion neutron source.

Section 26.3 contains an overview of the SABR design. Section 26.4 describes the calculational model used to simulate transients in SABR. Section 26.5 contains the results of the transient simulations of accidents initiated in SABR by a variety of events, and finally, Section 26.6 includes a summary and conclusions about the safety characteristics of SABR.

26.3 SABR#1 design overview

SABR#1 is a loop-type sodium-cooled, metal-fueled, fast transmutation reactor with a variable strength tokamak D–T fusion neutron source. A simplified, three-dimensional view of the design is shown in Figure 26.1. Outside the toroidal tokamak fusion plasma chamber is a 62-cm-thick annular fission core. Both the fusion plasma and fission core are surrounded by reflector, tritium-breeding blanket, and shield regions—a combined thickness of 80 cm. Sixteen D-shaped superconducting toroidal field magnets surround both the fusion plasma chamber and fission core, as well as the reflector, breeding blanket, and shield. Figure 26.1 does not show the plasma's divertor system, the sodium coolant pipes, or the control rod drives.

SABR#1 is designed to transmute the TRU in SNF to reduce geological high-level-waste storage requirements, in the process producing 3,000 MW(thermal), which is then converted into electricity. SABR#1's 3.2-m-tall fission core is composed of 2 m of active fuel, a 1-m fission gas plenum, and a 20-cm reflector capping the fuel pin. There are 271 fuel pins in each of the 918 hexagonal fuel assemblies, which are arranged vertically in four annular rings on the outboard of the plasma chamber. Each pin is 4 mm in diameter and composed of transuranic-zirconium fuel and clad in oxide dispersion-strengthened (ODS) steel. In addition to the 902 fuel assemblies, there are also 16 control rod assemblies composed of boron carbide that provide about 9$ of negative reactivity.

SABR#1's metallic fuel has an initial weight percent composition of 40Zr-10Am-10Np-40Pu. This fuel composition, which is under development at Argonne National Laboratory, was chosen because of its high thermal conductivity, high fission gas

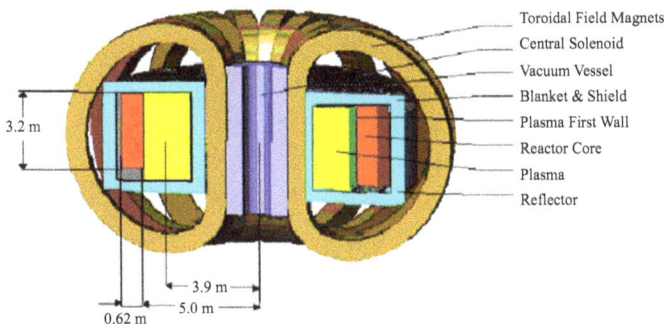

Figure 26.1   SABR#1 configuration

retention, and ability to accommodate high actinide density. Zirconium was added to the fuel to create a small negative Doppler feedback coefficient, to provide stability during irradiation, and also to increase the melting temperature of the alloy to 1,473 K.

SABR's fusion neutron source is capable of generating up to 500 MW. Table 26.1 lists the required fusion power levels for the fission core to maintain 3,000 MW(thermal) at BOL, BOC, and EOC as well as the effective multiplication constant of neutrons born in the fission core, k_{eff}, and the multiplication constant in the fission core of the neutrons produced in the fusion neutron source, k_m. Neutron multiplication levels for SABR are from the fuel cycle calculations in [6].

To cool the reactor, a three-loop cooling system is utilized with sodium in the primary and intermediate loops and water in the secondary loop. Heat generated in the fission core is transferred through four intermediate straight-tube heat exchangers to sodium in the intermediate loop. The hot sodium in the intermediate loop then converts water in the secondary loop to steam, which then passes through turbines to generate electricity. Sodium in the primary loop flows through the core at 8,695 kg/s in a total flow area of 7.5 m². At steady state the inlet temperature to the core is 650 K, and the outlet temperature is 923 K. Because it is difficult to adequately mix all of the sodium prior to the inlet plenum, the sodium travels through two separate circuits in the primary loop. Each circuit has four electromagnetic, EM, coolant pumps and two IHXs for a total of eight primary coolant pumps and four IHXs. (We assume that these pumps would be connected to different power sources to protect against multiple pump failures.) Figure 26.2 illustrates the coolant flow

Table 26.1   Required fusion neutron source strength versus time in fuel cycle (BOL is beginning of life, BOC is beginning of cycle, EOC is end of cycle)

	BOL	BOC	EOC
k_{eff}	0.972	0.894	0.868
Reactivity -$!	−9.26	−35.2	−43.8
k_m	0.913	0.753	0.676
P_{fus}, MW	73	242	370

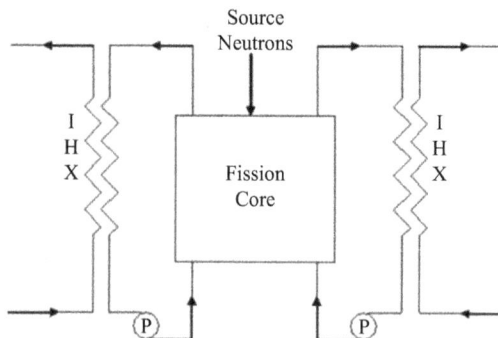

Figure 26.2   Sodium heat removal system for SABR

path that was used in the RELAP5-3D calculational model, with each pump in the figure representing four pumps and each heat exchanger representing two heat exchangers.

26.4   Dynamics calculation model

Transients develop differently in critical and subcritical systems. In a critical system, a small introduction of positive reactivity will lead to a prompt jump of the fission power followed by a gradual power increase due to the increased delayed neutrons. However, positive reactivity insertions larger than the delayed neutron fraction will make the reactor prompt supercritical, leading to a rapid exponential power increase. Negative reactivity insertions in a critical system will cause a prompt drop in fission rate followed by a further decay of the fission rate on the delayed neutron timescale. Negative reactivity feedback mechanisms can counter these reactivity changes and bring the reactor to critical at a different power level, and control rods can be used if these feedback effects are insufficient.

Reactivity insertions in subcritical systems, however, will always lead to new steady-state power levels as long as the neutron source remains active and the reactor remains subcritical. No matter how large the negative reactivity insertion, the reactor will continue to generate some power in the presence of a neutron source. Shutting off the neutron source is a quick and effective way to reduce the fission core power level in a subcritical core to decay heat levels, we have not identified an inherent negative feedback mechanism for shutting off the neutron source during a fission power excursion, but this search should continue.

Dynamics calculations for SABR during various transient conditions were performed using the RELAP5-3D thermal-hydraulic code, which is able to couple the power generation of the fission core with the thermal hydraulics of the heat removal system. The ATHENA version of RELAP5-3D allows for the simulation of liquid-metal coolants such as sodium. Through a series of connected one-dimensional volumes and junctions, the hydrodynamic state of the sodium coolant can be calculated at all points in the system. Using the point kinetics neutronics equations, RELAP5-3D is able to calculate the total power level of the reactor, which determines how much power is generated in the fuel pins and how much heat is transferred to the coolant. Use of point kinetics equations is a good approximation for SABR's fission power level because of the high level of subcriticality at which SABR operates, which in addition to an annularly symmetric neutron source, should lead to an annularly symmetric dynamic response of the neutron flux in the fission core during transient scenarios. We use the Doppler and Na-voiding coefficients given in Table 26.2.

RELAP5-3D calculations using peak and average power fuel assemblies allows for the calculation of the maximum and average axial and radial temperature distribution of the fuel pins. The conditions of the reactor coolant throughout the

Table 26.2   Reactivity feedback coefficients

	BOL	**BOC**	**EOC**
Doppler dk/dT_{fuel}	$-2.32\text{E}-8^{\text{a}}$	$-8.62\text{E}-7$	$-9.81\text{E}-8$
Sodium voiding dk/dT_{cool}	$6.01\text{E}-6$	$2.78\text{E}-5$	$8.87\text{E}-6$

[a]Read as -2.32×10^{-8}.

transients can be tracked through the core, primary coolant loop piping, IHX, and other locations.

C_j = delayed neutron precursor population density
b_j = fraction of fission neutrons resulting in the formation of a delayed neutron precursor in group "j"
λ_j = delayed neutron precursor decay constant
n = neutron density in the reactor
L = neutron lifetime in the reactor.

The total reactivity ρ_{total} is the sum of the subcritical reactivity, the reactivity due to feedbacks, and any external reactivity insertions such as control rod withdrawal. Changes in the density of the coolant as well as the temperature of the fuel pins can be used to calculate the reactivity feedbacks. SABR should have a substantial negative fuel expansion coefficient, but making an accurate calculation of this was beyond the scope of this initial effort, and it was not included in the calculations reported in this paper. The total reactivity for the neutron kinetics calculation was

$$\rho_{\text{total}} = \rho_{\text{sub}} + \rho_{\text{Doppler}} + \rho_{\text{Na-void}} + \rho_{\text{expansion}} + \rho_{\text{external}}$$

The fusion neutron source is incorporated into the calculations through the source density term S in the point kinetics equations. RELAP5-3D calculates the necessary source strength required for steady-state operation at the beginning of the simulation. Unfortunately, this number cannot be changed during the RELAP5-3D simulation to represent changes in the neutron source strength. However, the effect of changes in the source strength can be simulated indirectly by calculating fictitious changes in the external reactivity that would produce the same change in neutron density.

The power distribution in the fission core throughout the transient simulations has been simulated by a 33-group ERANOS calculation and represented in the coolant loops in the system. Heat generated in the fuel pins is calculated as a function of the core power level and then transported across the fuel pins into the coolant and circulated to the heat exchanger coolant pipes, etc. Figure 26.2 illustrates the coolant flow path that was used for the RELAP5-3D calculations. The two coolant flow paths represent the two separate circuits in the primary coolant loop. Each pump in the figure represents four EM coolant pumps while each heat exchanger represents two IHXs in the RELAP5-3D model.

Axial and radial power peaking factors for the fission core were generated using the ERANOS code system. The axial pin power distribution was assumed to be sinusoidal. Reactivity feedback coefficients for Doppler and sodium voiding were also generated using ERANOS and are given in Table 26.2. The Doppler coefficients are negative but very small. The positive BOL and EOC sodium voiding worths are small but positive. The negative reactivity feedback associated with axial core expansion in metallic-fueled cores due to increased core temperatures during accident scenarios was not included in this study but should be considered during future analyses of SABR's response to transients. The delayed neutron parameters were found using the VAREX variational analysis code and are given in Table 26.3.

26.5 Accident simulations

Two types of accidents are simulated to determine the dynamic safety characteristics of SABR. Accidents affecting SABR's heat removal capability in the fission core are the first type

Loss-of-flow accident (LOFA)
Loss-of-heat sink accident (LOHSA)
loss-of-power accident (LOPA).

Accidents that affect the neutron population of the fission core are the second type

accidental reactivity insertions;
accidental increases in fusion-generated source neutrons.

Simulations of the fusion neutron source indicate that simply turning off the plasma heating power source will cause the fusion neutron source to be reduced to a negligible level within a few seconds on the energy confinement timescale. This type of reduction in neutron source strength reduces the power in the reactor core to the decay heat level in a few seconds.

We have not attempted to simulate detection of accidents and shutdown of the neutron source to drop the power level quickly to decay heat levels. Rather, we have assumed that the transients are undetected and left the neutron source on except, of

Table 26.3 Delayed neutron parameters

Group	β_i	λ_l (1/s)
1	8.308E−5[a]	1.324E−2
2	7.623E−4	3.019E−2
3	5.836E−4	1.166E−1
4	1.059E−3	3.133E−1
5	4.139E−4	1.046
6	1.069E−4	2.837
Total	3.009E−3	—

[a]Read as 8.308×10^{-5}.

course, in the LOPA. The results give us insight into the consequences of undetected/uncorrected accidents and also provide an indication of how much time is available for detection and correction before core damage occurs. Because RELAP5-3D does not allow a time-dependent neutron source, we were unable to calculate the time evolution of those accidents involving inadvertent transients in the plasma power output, but we were able to calculate the bounds of such accidents. Accidents were simulated at three different points during the fuel cycle: BOL, BOC, and EOC.

The main criterion used to determine if the core had reached a limiting condition during an accident is whether the fuel had exceeded its melting temperature of 1,473 K at any location, which would cause core damage. A secondary failure criterion that was monitored was whether the coolant had exceeded its boiling temperature of 1,156 K. If the coolant begins to boil during a transient, this does not necessarily mean that core damage will occur, because corrective measures could lead to decreased core temperatures and the sodium vapor would return to its liquid phase. On the other hand, fuel melting cannot be reversed and will lead to significant down time and repairs for the reactor, if not permanent shutdown. During the initial scoping simulations, it was found that the sodium boiled before the fuel melted in all accidents that reached limiting conditions.

Melting of the ODS cladding is not considered because the 1,800 K melting temperature of ODS steel is far greater than the 1,473 K melting temperature of the fuel, which will always be hotter than the cladding. Cladding creep is also not considered because it occurs at relatively high temperatures and is only an issue if the clad remains at elevated temperatures for extended periods of time. Because damage to the core due to eutectic melting between the cladding and the fuel is not well understood and, as with the cladding creep, occurs at elevated temperatures over longer periods of time, the eutectic limit is not considered as a constraint in the accident scenarios. To fully understand how different transients affect SABR, it is useful first to look at SABR's typical operating parameters during steady state. Table 26.4 lists some of the important temperatures in the fission core during normal 3,000-MW operation at the three reference points during the fuel cycle.

Because of the SABR core's relatively thin annular geometry and the presence of a neutron source on the inside, the fission core will experience higher radial

Table 26.4   Steady-state operating parameters for SABR calculations

	BOL	**BOC**	**EOC**
Coolant inlet temperature, K	650	650	650
Average coolant outlet temperature, K	942	942	942
Maximum coolant temp, K	942	942	971
Maximum cladding temperature, K	957	956	986
Maximum fuel temperature, K	1,006	1,048	1,087
Average coolant temperature, K	815	815	815
Average cladding temperature, K	824	824	824
Average fuel temperature, K	838	838	838
Radial power peaking	1.26	1.80	1.96

power peaking values than normally found in fast reactors. Special attention must be given to the coolant inlet channels to ensure that the outlet temperature of the hot assemblies is not too close to the coolant boiling temperature. During BOL and BOC operation, it is possible to achieve similar outlet temperatures for both hot and average assemblies by modifying the channel inlet flow conditions to provide more coolant mass flow to the necessary assemblies. Because the fuel is not shuffled between BOC and EOC operation, care must be taken when selecting the flow inlet conditions for BOC operation to ensure that the hot assembly outlet temperature at EOC is not too high.

The preliminary safety and dynamics analyses for SABR indicated that the best way to effect small changes to the power level in the fission core is with control rods, while fully shutting down the reactor can only be accomplished by turning off the neutron source.

The plasma power balance is maintained by an external source of heating power, which can be switched off essentially instantaneously, after which the plasma cools on the energy confinement timescale, which is a few seconds. Turning off the plasma auxiliary heating for the neutron source will result in a negligible neutron source in 3 s, which would cause the fission power level to promptly decrease to decay heat levels, or 7.1% of the steady-state power level. *Turning off the plasma heating power serves as an excellent rapid scram system for a sub-critical reactor driven by a fusion neutron source.*

26.5.1   Loss-of-flow accident

A LOFA, often due to pump failure, is an accident in which the reactor core experiences less coolant mass flow, resulting in inadequate heat removal of the heat generated in the fuel pins, leading to significant increases in the fuel temperature. For a uranium-fueled reactor with a large negative Doppler feedback, this increase in fuel temperature will lead to a significant negative reactivity insertion and a consequent decrease in fission power. However, the Doppler coefficient in a TRU-fueled transmutation reactor such as SABR#1 is small. During a LOFA, if neither coolant nor fuel failure occurs, the reactor will reach a new equilibrium where the power produced in the core is removed in the heat exchanger.

A significant problem encountered when designing any reactor is ensuring that a decrease in coolant mass flow is distributed over the whole core and not over just a few assemblies. The problem is getting the coolant to mix in the structure and piping below the active region of the core so that if one coolant loop experiences decreased flow, the change is spread over all the assemblies. This problem is more difficult in SABR#1 because of the annular nature of the core and the presence of the neutron source that prevents a cylindrical mixing area below the core.

LOFAs were simulated as the complete failure of one or more of the four primary loop pumps, while the neutron source remains active. These simulations assume that the coolant leaving the four pumps in each half of the reactor is adequately mixed before flowing back through the core. Control rods can be used to decrease the core fission power level, but as long as the neutron source remains on, there will be significant fission power production. Shutting off the neutron source is the best and perhaps only way to counteract the effects of a LOFA without the use of auxiliary pumps or restarting the failed pumps. This emphasizes the importance of connecting different pumps to different power sources.

A series of pump failure transients have been simulated without control action to turn off the neutron source to determine the amount of time that would be available to detect the accident and shut down the neutron source before core damage occurred. For an accident that results from the failure of a single primary loop coolant pump, SABR can survive without sustaining fuel melting or coolant boiling during all reference points during the fuel cycle—BOL, BOC, and EOC. The highest coolant and fuel pin temperatures, 1,068 and 1,172 K, respectively, were experienced at EOC, but these temperatures were far below the coolant boiling and fuel melting temperatures of 1,156 and 1,473 K, respectively.

If two or more pumps on the same side of the primary loop were to fail, however, the coolant mass flow rate would decrease too much to adequately remove the heat generated in the fuel pins. During BOL operation, there would be at most 13.2 s after the loss of the two pumps before the onset of coolant boiling. This time drops to 7.1 s for the loss of two pumps during EOC operation. The decreased time before coolant boiling at EOC is due to the higher outlet temperature for the hot assemblies during steady-state EOC operation. Fuel melting did not occur after the failure of two coolant pumps on the same side of the primary loop. Figure 26.3 illustrates the maximum coolant and fuel temperatures during a 25% and a 50% LOFA. Fuel melting was only experienced 15.5 s after three of the four coolant pumps on the same side of the primary loop failed at BOL. There are only 9.7 and 8.4 s after the start of the transient at BOC and EOC, respectively, before fuel melting occurs. At this point SABR would have sustained irreversible damage to the core. Because of the highly subcritical nature of the reactor, increases in the fission power due to the positive sodium void effect during LOFAs

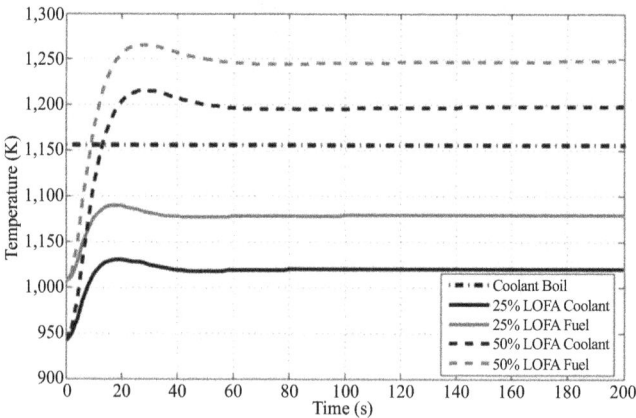

Figure 26.3   Maximum coolant and fuel pin temperature during 25% and 50% LOFA at BOL with fusion neutron source remaining on

were not large. For example, a 50% LOFA during BOL resulted in a fission power increase of only 22 MW. *We note again that the likelihood of this multiple pump failure in the same loop can be minimized by connecting different pumps to different power supplies.*

Because SABR cannot withstand multiple pump failures without at least the coolant boiling, the pumps in the primary loop should be kept on separate systems so that a failure of one pump is unlikely to be followed by a second pump failure. Once a single pump failure is detected, most likely by the detection of decreased coolant mass flow, the fusion neutron source should be turned off or reduced until the problem can be corrected. The remaining pumps would be able to provide enough coolant mass flow so that the broken pump can be isolated and repaired. Table 26.5 summarizes the maximum temperatures reached during these accidents and the time from the start of the accident until coolant boiling or fuel melting occurs.

Future work performed on designing SABR's heat removal system must ensure that a loss of pumping power in a coolant loop does not lead to stagnant coolant in any one reactor region.

26.5.2   Loss-of-heat-sink accident

A LOHSA is any transient that results in an unanticipated decrease in the overall heat transfer rate of the reactor.

This decrease could be a result of a line break in the intermediate loop, decreased coolant flow in the intermediate loop, or a physical break in the intermediate hear exchanger (IHX). Without adequate heat removal, primary loop temperatures will continue to increase until either a new steady state is reached or

Table 26.5   Loss-of-flow accident summary

	BOL	**BOC**	**EOC**
25% LOFA			
Maximum coolant temperature, K	1,030	1,032	1,068
Maximum fuel temperature, K	1,090	1,126	1,172
Time until coolant boiling, s	–	–	–
Time until fuel melting, s	–	–	–
50% LOFA			
Maximum coolant temperature, K	1,216	1,217	1,266
Maximum fuel temperature, K	1,266	1,302	1,360
Time until coolant boiling, s	13.2	9.6	7.1
Time until fuel melting, s	–	–	–
75% LOFA			
Maximum coolant temperature, K	>1,156	>1,156	>1,156
Maximum fuel temperature, K	>1,473	>1,473	>1,473
Time until coolant boiling, s	8.4	5.7	4.9
Time until fuel melting, s	15.5	9.7	8.4

Note: Coolant boiling occurs at 1,156 K and fuel melting occurs at 1,473 K.

core failure is experienced. SABR's positive sodium voiding reactivity feedback will lead to a further increase in the fission power.

Because simulating a physical break of the heat exchanger is difficult, the decrease in the IHX heat transfer rate was represented by a decrease in the coolant mass flow of the intermediate loop as a result of a failure of one or more coolant pumps in the intermediate loop. *As with other accidents, shutting down the neutron source is the best and fastest way to prevent core damage due to a LOHSA.* The accidents were simulated without any control action to determine the amount of time available to detect the accident and shut down the neutron source.

At all three reference points—BOL, BOC, and EOC—in the fuel cycle, SABR can withstand up to a 50% LOHSA without coolant boiling or fuel melting, which corresponds to half of the intermediate loop coolant pumps failing. At EOC, the maximum coolant temperature reaches 1,132 K, which is close to the sodium boiling temperature of 1,156 K.

Anything greater than a 50% LOHSA will ultimately lead to both coolant boiling and fuel melting. For a 75% LOHSA, the least amount of time before the failure criteria are met is during EOC when there will be 44.9 and 146.6 s before the coolant boils and the fuel melts, respectively. During a complete LOHSA, those times drop to 24.2 and 56.5 s. As with the primary loop pumps, despite the fact that the reactor can survive more than one intermediate loop coolant pump failing, the pumps should be kept on separate systems so that the likelihood of more than one pump failing is minimal. The maximum coolant and fuel temperatures during various LOHSA are illustrated in Figure 26.4. The results of these accidents are summarized in Table 26.6.

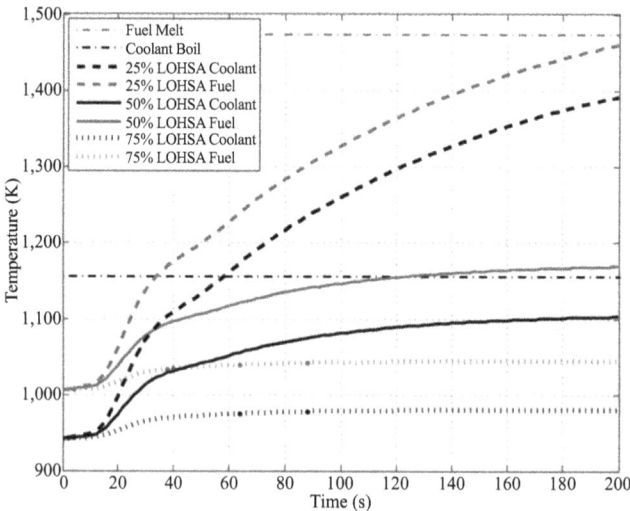

Figure 26.4   Maximum coolant and fuel pin temperatures during 25%, 50%, and 75% LOHSA at BOL with neutron source remaining on

Table 26.6 Loss-of-heat-sink accident summary

	BOL	**BOC**	**EOC**
25% LOHSA			
Maximum coolant temperature, K	981	981	1,009
Maximum fuel temperature, K	1,045	1,080	1,118
Time until coolant boiling, s	∞	∞	∞
Time until fuel melting, s	∞	∞	∞
50% LOHSA			
Maximum coolant temperature, K	1,107	1,108	1,132
Maximum fuel temperature, K	1,173	1,208	1,243
Time until coolant boiling, s	∞	∞	∞
Time until fuel melting, s	∞	∞	∞
75% LOHSA			
Maximum coolant temperature, K	1,156	1,156	1,156
Maximum fuel temperature, K	1,473	1,473	1,473
Time until coolant boiling, s	58.6	54.1	44.9
Time until fuel melting, s	217.7	170.7	146.6
100% LOHSA			
Maximum coolant temperature, K	1,156	1,156	1,156
Maximum fuel temperature, K	1,473	1,473	1,473
Time until coolant boiling, s	29.7	26.4	24.2
Time until fuel melting, s	70.6	61.0	56.5

Note: Coolant boiling occurs at 1,156 K and fuel melting occurs at 1,473 K.

26.5.3 Loss-of-power accident

A LOPA is an accident where power to SABR's auxiliary systems is lost, meaning all coolant pumps turn off. Heating power to the plasma also turns off. This accident is similar to a complete LOFA with immediate corrective measures. When plasma electrical power is lost, the neutron source will quickly shut down, as discussed above, leaving a subcritical fission core without the source neutrons necessary to remain at steady state. The power level in the fission core will quickly decrease to decay heat levels. Decay heat production in the core is initially 212 MW(thermal), or 7.1% of the steady-state power level. Before coolant mass flow has a chance to significantly reduce, the core will quickly cool down. But, as the coolant mass flow decays away, the core will begin to heat back up. If natural circulation does not provide enough coolant mass flow, the core will overheat and eventually the coolant will boil and the fuel will melt.

It is clear that decay heat removal systems should be included in *future* designs. Because the source shuts off so quickly in SABR during a LOPA, the fission power is able to decay away faster than the coolant mass flow, leading to only a small increase in core temperatures. The maximum coolant temperature reached during BOL is 1,054 K while the maximum fuel pin temperature reaches 1,057 K, both an acceptable margin below their respective failure points of 1,156 and 1,473 K. The BOC and EOC maximum coolant temperatures are 1,103 and 1,130 K, respectively, which are closer to but still below the sodium boiling

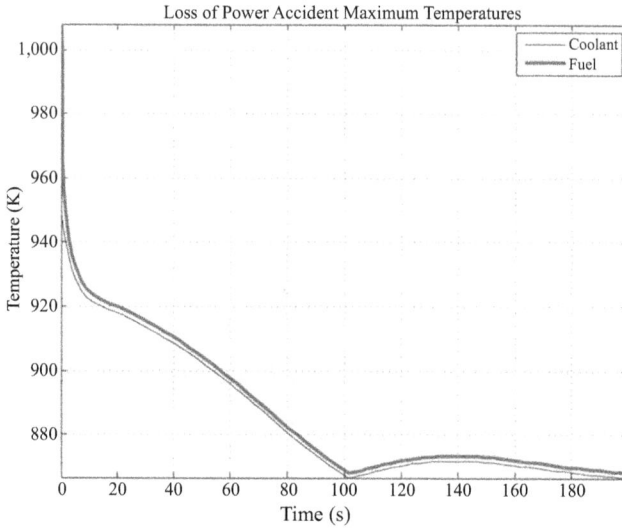

Figure 26.5   Maximum fuel and coolant temperatures during LOPA at BOL without neutron source

Table 26.7   Loss-of-power accident summary

	BOL	BOC	EOC
Maximum coolant temperature, K	947	1,103	1,068
Maximum fuel pin temperature, K	1,008	1,106	1,083

temperature. The maximum temperatures for a LOPA during EOC operation are higher because of the higher steady-state hot assembly temperatures.

Even though the coolant during a LOPA reaches a maximum temperature close to the coolant boiling temperature, core temperatures quickly decrease until natural circulation begins to cool the decay heat produced in the reactor. Ten minutes after the LOPA is initiated, core temperatures are all <850 K, and they continue to decrease. The maximum coolant and fuel temperatures during a LOPA are illustrated in Figure 26.5. The results of this accident are summarized in Table 26.7.

26.5.4   Worst possible control rod accident

In a critical system, even the smallest reactivity insertion can lead to a steady power increase in the absence of control mechanisms or negative feedback. Reactivity insertions in a subcritical system will, however, always lead to a new steady-state power level as long as the reactor remains subcritical. In SABR, the most reactive condition occurs at BOL with entirely fresh fuel in the reactor. At this point the reactor is negative 9.26\$ subcritical, and k_{eff} is 0.972. It would require an enormous positive reactivity insertion to reach criticality.

If during BOL operation the 16 control rods—which are initially withdrawn and are for intended safety purposes, not compensation for reactivity changes—were fully inserted, the reactor would be operating at negative 18.26 subcritical. In order to maintain 3,000 MW(thermal), the fusion neutron source strength would have to be nearly doubled by the operator. If after the fusion neutron source strength was doubled, the control rods were fully removed and $k_{eff} = 0.972$ reestablished, the new steady-state power level in the fission core would be 5,884 MW(thermal). The peak fuel temperature would increase to 1,345 K, still below the melting temperature of 1,473 K, but the coolant temperature would exceed its boiling point of 1,156 K in 13 s. SABR could withstand this accident without experiencing fuel melting; however, the occurrence of sodium boiling in the fission core necessitates that changes be made to the reactor design to minimize the magnitude of this accident. It should be noted that this accident is highly implausible, but it was simulated because it is the worst possible accident related to reactivity insertions that we could imagine.

One solution is to decrease the worth of the control rods. Because large negative reactivity insertions from the control rods are not necessary to control or shut down the reactor, a lower total control rod worth could be utilized to eliminate the potential of damage from this accident scenario. Past studies of SABR's safety characteristics concluded that a total control rod worth of approximately half the BOL subcritical reactivity would result in maximum coolant temperatures just below boiling during this accident.

Another method to prevent the possibility of damage from this accident is to decrease the BOL effective multiplication constant. The original SABR design used a maximum value for k_{eff} of 0.95, as compared to the current value of 0.972, and was able to withstand this accident without the occurrence of either fuel melting or coolant boiling. Future iterations of SABR will likely utilize both methods to provide a significant safety margin against damage as a result of this highly implausible accident. The reactor could be made entirely safe from damage. A summary of the temperature and fission power increases for this accident is given in Table 26.8.

Table 26.8 Summary of reactivity insertion transients

	BOL	BOC	EOC
Control rods inserted, fusion power increased to compensate, control rods removed			
Maximum coolant temperature, K	>1,156	1,015	1,035
Maximum fuel temperature, K	1,377	1,148	1,173
Time to coolant boiling, s	12.6	,	,
Time to fuel melting, s	,	,	,
Fission power increase, MW	2,884	754	600
Maximum coolant temperature, K	963	947	975
Maximum fuel temperature, K	1,031	1,055	1,093
Fission power increase, MW	190	48	38

Note: Coolant boiling occurs at 1,156 K and fuel melting occurs at 1,473 K.

26.5.5   Control rod ejection

A more likely reactivity insertion would be the ejection of one control rod, which corresponds to a reactivity insertion of 0.56$. At BOL this would lead to an increase in fission power of only 190 MW. The peak coolant and fuel pin temperatures increase by only 21 and 25 K, respectively. SABR could withstand the inadvertent ejection of a control rod from the core without sustaining permanent damage to the reactor.

26.5.6   Accidental increase in fusion neutron source strength

In addition to increasing the reactivity in the fission core, the neutron population, thus the fission power level, also can be increased by increasing the strength of the fusion neutron source. Note that the total neutron source in the reactor consists of three terms: prompt fission neutrons, delayed fission neutrons from the decay of delayed neutron precursors, and the fusion source neutrons.

It is possible, although unlikely, that an accidental increase in the plasma fueling source or the plasma heating source could cause an unintended increase in the plasma density or temperature, respectively, leading to an increase in the fusion power level and the fusion neutron source strength in the reactor core. It is also possible that changes in other plasma parameters that affect the particle and energy confinement, thus the fusion rate, could occur. Any change in plasma density or temperature would take place on the plasma confinement timescale of a few seconds.

As a concrete example, the SABR fusion neutron source has available six 20-MW lower hybrid EM wave launchers to provide auxiliary heating to the plasma. Not all of these launchers are needed for all conditions in the fuel cycle, so there is the possibility that an unused launcher is inadvertently turned on. In a similar vein, there are gas injectors for fueling the plasma that are not always needed, and one of these could inadvertently be opened.

Just as there are negative temperature feedback mechanisms that act to reduce the reactivity and thereby limit power excursions in fission reactors, there are also negative density and temperature feedback mechanisms—density limits and pressure beta limits—that would act to reduce the energy and particle confinement and thereby limit power excursions in the fusion neutron source. As the density approaches the density limit or the pressure approaches the beta limit, the plasma confinement decreases strongly, thereby limiting positive excursions in density or temperature, provided that the plasma operating conditions are chosen sufficiently close to these limits. No attempt has yet been made to optimize the SABR neutron source operating point with respect to these density and pressure limits.

Because RELAP5-3D uses a constant value for the source term in the point kinetics equations, it is impossible to capture all of the dynamics of the fusion neutron source for these calculations. For this reason, a series of calculations was first performed using the point kinetics equations to represent the fission power level, the equivalent global plasma power and particle balance equations to represent the fusion neutron source dynamics, and a simple model for the sodium heat removal system. The density and pressure limits were represented by empirical expressions but were not incorporated into the dynamics calculations. Rather, when one of these limits was surpassed it was assumed that the neutron source had been turned off by plasma confinement degradation.

The results of these calculations provided useful insights that guided subsequent RELAP5-3D calculations, but the quantitative results were questionable because of the simplified heat removal system model employed. To obtain a quantitative estimate of limiting temperatures and power levels in the fission core that might occur as a result of fusion source excursions, a second series of calculations was carried out with the more realistic RELAP5-3D model of the heat removal systems, using fixed plasma sources at various levels to determine the maximum fusion neutron source strength that SABR could tolerate without experiencing coolant boiling or fuel melting in the fission core.

The Greenwald density limit is a simple empirical fit that bounds the stable densities for tokamak plasmas. The plasma neutron source operating density was well below this limit and was not reached in any of the transients simulated.

The Troyon beta limit provides an upper limit on both the plasma temperature and the ion density, beyond which plasma stability and confinement are degraded. The plasma-operating parameters used during BOL operation provided a large margin at steady state before the Troyon beta limit was exceeded. However, the margin during BOC and EOC operation before the Troyon beta limit was exceeded was much smaller. It was apparent that optimization of the plasma current profile was necessary for BOC and EOC plasma-operating parameters to allow for operation just enough below this limit to provide an inherent feedback against power excursions in the fusion neutron source due to inadvertent turn on of additional plasma heat sources.

Simulations representing inadvertent increases in plasma fuel injection indicated that at BOL, BOC, and EOC, SABR could withstand up to 11%, 1%, and 2% increases, respectively, in the plasma ion density before the Troyon beta limit was exceeded. When compared with the 12%, 17%, and 19% increases that were required before coolant boiling occurred, it was determined that the Troyon beta limit provided a natural feedback mechanism that prevented reactor damage due to accidental increases in the plasma fueling rate. Ion density increases of 19%, 29%, and 32% at BOL, BOC, and EOC, respectively, were required before fuel melting occurred.

It should be noted that changes in the plasma ion density did not produce a linearly proportional change in the fusion power level. For a 5% increase in the plasma ion density at BOL, the fusion power increased by more than 20%. A 12% increase in the plasma ion density at BOL led to a nearly 60% increase in the fusion power level.

The second fusion transient that was examined for the fusion neutron source was an increase in the plasma auxiliary heating. This could result from one or more unused 20-MW heating launchers accidentally turning on. For the case of one extra 20-MW plasma auxiliary heating launcher turning on at BOL, SABR experienced a 41% increase in fusion power and maximum coolant and fuel temperatures of 1,079 and 1,142 K, respectively, both below their respective failure temperatures of 1,156 and 1,473 K. In every other case considered, whether it was two or more extra 20-MW auxiliary heating launchers or one extra 20-MW heating launcher at BOC and EOC, the Troyon beta limit was exceeded, indicating that the plasma energy confinement would have decreased and terminated the source excursion before any damage to SABR's fission core occurred.

Because more complicated transients for the fusion neutron source could not be simulated with the RELAP5-3D model, the maximum allowable neutron source strength was calculated before either coolant boiling or fuel melting occurred. Changes in the neutron source strength may be simulated as external reactivity insertions when using the point kinetics equations. This is a valid approximation because both the delayed neutron fraction β and the neutron lifetime l will change a negligible amount during changes in either the level of subcriticality or the neutron source strength. The response due to reactivity feedbacks was allowed to progress as it naturally would so that the final fission power level due to a change in the neutron source strength was maintained. Calculations were performed to determine what the necessary external reactivity insertion would be to match various neutron source strength changes.

The various tables list the minimum fusion power levels that led to coolant boiling in the fission core. In all three cases the fusion power level must increase by more than 50% before coolant boiling occurs. The corresponding maximum fuel temperatures calculated for an increase in the fusion neutron source power level at BOL, BOC, and EOC were all 100 K below the fuel melting temperature of 1,473 K.

26.6   Summary and conclusions for "point kinetics" analyses

The LOPA, LOHSA, LOFA, control rod ejection, and neutron source excursion accidents for the SABR#1 loop-type fast reactor were simulated using a point kinetics neutronics model and the RELAP5-3D thermal-hydraulics model, with a single set of plasma particle and power balance equations. Other types of accidents were also analyzed. The general conclusion was that the reactor could survive a loss coolant flow up to about 50% without damage.

There is a more efficient way than burial to dispose of SNF and the long-lived TRUs contained within. TRU can be destroyed by neutron fission in a fast burner reactor. The high-energy neutron spectrum of a fast reactor enables it to fission TRU and transmute them into shorter-lived fission products. This process destroys TRU and produces more recoverable energy from the uranium fuel at the same time.

Critical fast burner reactors require uranium driver fuel to be mixed in with the TRU in order to increase the reactivity margin to prompt critical above the delayed neutron fraction of TRU ($\beta \approx 0.002$). The requirement for uranium fuel reduces the amount of TRU burned in the reactor, and it also produces some TRU as well. This reduces the efficiency of critical fast burner reactors at burning TRU.

A more efficient way is to have a reactor completely fueled with TRU operated subcritical with an adjustable neutron source used to maintain the neutron chain reaction.

There are two types of external neutron sources that have been suggested for this. The first is an accelerator-driven system in which accelerated deuterons impinge on a spallation neutron target embedded in the TRU fuel assemblies within the reactor. The second type is a fission–fusion hybrid reactor in which the 14.1-MeV neutrons from deuterium–tritium fusion reactions stream into TRU fuel assemblies blanketing the outside of the plasma chamber.

The subcriticality of a source-driven system provides two distinct advantages. As the fissionable TRU are depleted by burnup and their reactivity decreases, the fission chain reaction can be maintained at the same power by increasing the neutron source, resulting in increased fuel burnup to the radiation damage limit before the fuel is removed from the reactor. The second advantage is an increased reactivity safety margin to a prompt critical power excursion. For a critical reactor, this safety margin is $\Delta\rho \approx \beta$, the delayed neutron fraction. On the other hand, for a subcritical reactor the safety margin is $\Delta\rho \approx \Delta k_{\text{sub}} = 1 - k \gg \beta$.

The SABR is a combination of an ITER-like fusion neutron source and an integral fast reactor (IFR)-like fast reactor. The annular fission reactor lies just outside the toroidal plasma chamber as shown in Figure 26.6, and it is composed of ten separate sodium pools each with its own fission core and heat exchanger. One

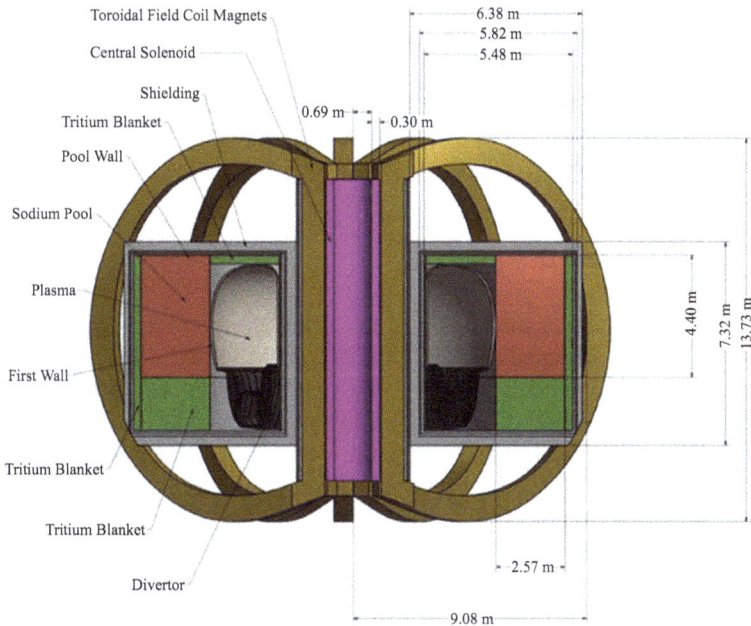

Figure 26.6   Overview of SABR#2

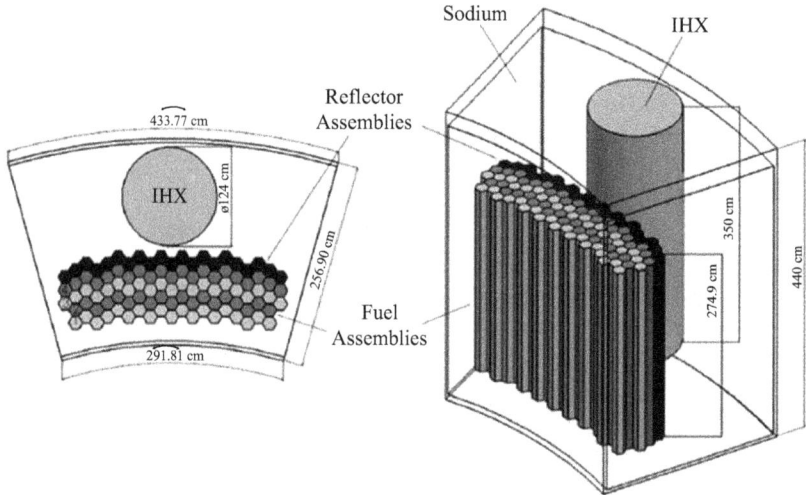

Figure 26.7   SABR#1 pool layout

of the pools is shown in Figure 26.7. Each pool contains a metal-fueled, sodium-cooled, pool-type, IFR-like fast reactor that produces 300 MW(thermal). A formal design paper on SABR is given in [12].

Argonne National Laboratory (ANL) designed IFR to be passively safe. SABR makes use of the same Na-pool, metal-fueled technology as IFR, but there are several design differences between SABR and IFR [17]. IFR is a single-core, critical reactor. SABR is a subcritical, source-driven reactor with ten physically separate but neutronically coupled cores. A change in the conditions in one core may affect the conditions of the neighboring cores in different ways. This chapter describes a dynamic safety analysis of SABR.

26.7   Nodal dynamics model

A nodal neutron dynamics model is used to calculate the evolution of the power in the ten neutronically coupled cores of SABR#2. COMSOL Multiphysics is used as the platform of the dynamics model [4]. COMSOL has 25 different physics packages available, several of which include solvers for fluid flow-through pipes, fluid and solid heat transfer, and structural mechanics. COMSOL can couple to other programs like MATLAB, which is used to couple the neutron kinetics calculations to the thermal-hydraulic calculations. Figure 26.8 shows the overall computational flow of our model.

The MATLAB nodal kinetics solver uses tables of neutron kinetics data precomputed by the MCNP Monte Carlo code. For each time step in a transient, COMSOL sends the pool and node temperatures to the MATLAB kinetics solver. The MATLAB kinetics solver uses those temperatures to determine the change in

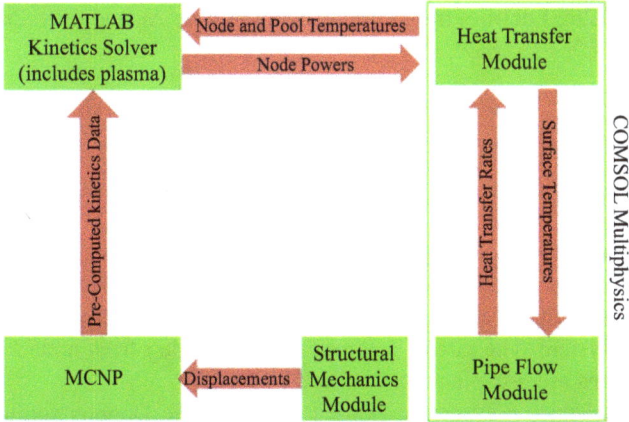

Figure 26.8   Computational flow of dynamics model

fission power levels for each node and sends them back to COMSOL. COMSOL then uses the new power levels to calculate the new node and pool temperatures for the next time step, etc.

26.7.1   Neutron kinetics model

The dynamics model uses a nodal neutron kinetics approach to calculate the time-dependent powers in each node (core) in SABR. The formulation is given in Eqs. (26.1) and (26.2):

$$\frac{dn_j}{dt} = \frac{\left(1 - \beta_j\right)}{\Lambda_{fj}} n_j(t) + \sum_{i=1}^{6} \lambda_{i,j} c_{i,j}(t) + \frac{n_j(t)}{\Lambda_{2nj}} + S_{fus,j} + \sum_{k=1}^{10} \frac{\alpha_{kj} n_k(t)}{l_{e,k}}$$

$$- \frac{n_j(t)}{l_{e,j}} - \frac{n_j(t)}{l_{a,j}} \tag{26.1}$$

and

$$\frac{dc_{i,j}(t)}{dt} = \frac{\beta_{i,j}}{\Lambda_j} n_j(t) - \lambda_{i,j} c_{i,j}(t) \tag{26.2}$$

$n_j(t)$ = neutron density of node j
β_j = delayed neutron fraction
$c_{i,j}(t), \lambda_{i,j}$ = six groups of delayed neutron precursor densities and their respective decay constants
$l_{e,j}, l_{a,j}$ = neutron leakage and absorption lifetimes, respectively
$S_{fus,j}$ = source neutrons contributed by the fusion plasma.

Equation (26.3) defines the fission generation time Λ_{fj} and the *n,2n* generation time Λ_{2nj}:

$$\Lambda_{fj} = \frac{1}{V\nu\Sigma_{f,j}} \quad \text{and} \quad \Lambda_{2nj} = \frac{1}{V\nu\Sigma_{n2n,j}}. \tag{26.3}$$

The *n,2n* reactions are a non-negligible source in the system due to the hard neutron spectrum.

A node k to node j coupling coefficient α_{kj} represents the neutronic coupling of the ten nodes; α_{kj} is the probability that a neutron emitted from the surface of node k will impinge upon the surface of node j before entering any other node or being lost from the system. The rate of neutrons entering node j from node k is α_{kj} times the rate neutrons are escaping from node k, $(n_k(t)/l_{e,k})$.

This model does not contain explicit terms for the Doppler/sodium/fuel/clad reactivity coefficients. Instead, the terms Λ_{fj}, Λ_{2nj}, $l_{e,j}$, $l_{a,j}$, $S_{\text{fus},j}$, and α_{kj} collectively represent the Doppler/sodium/fuel/clad reactivity coefficients. How these terms are calculated is discussed in Sections 26.7.2 and 26.7.3.

The nodal kinetics model constitutes a set of 70 coupled, stiff differential equations. We use the MATLAB program NodalSolve to numerically solve this system of equations for a given time step. NodalSolve uses MATLAB's intrinsic Ode15s solver, which efficiently solves ordinary, stiff differential equations.

26.7.2   Calculation of nodal kinetics terms

MCNP6 calculates each of the nodal kinetics terms using a three-dimensional (3D) model of SABR. Figure 26.9 shows cross sections of the 3D model. The model contains ten sodium pools in addition to ITER-like components such as the vacuum vessel, plasma chamber, divertor, breeding blanket, and central solenoid. The ITER-like components outside the pools are homogenized for simplicity. Inside the pools, the individual subcomponents are homogenized separately. The heat exchangers are represented as homogenized cylinders. The pool vessels are modeled as hollow trapezoidal prisms. Any space that is inside the pool vessels but

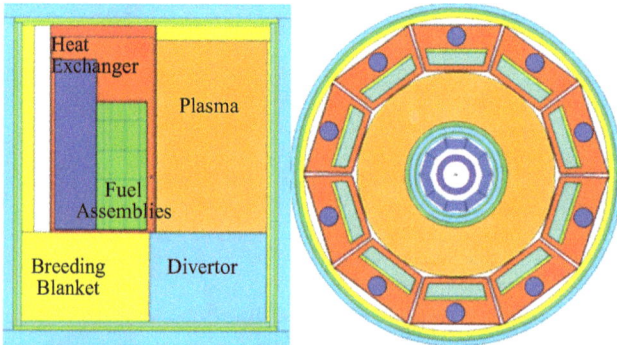

Figure 26.9   Side and top-down views of SABR 3D MCNP model

outside the cores and heat exchangers is by definition sodium. Each of the cores is divided into 125 cells and then homogenized. It is important to note that the cores are subdivided before being homogenized, not after. After the cores are subdivided and homogenized, they are referred to as nodes.

Different source and tally definitions are used depending on which kinetics terms are being calculated. To calculate Λ_{fj}, Λ_{2nj}, $l_{e,j}$, and $l_{a,j}$, an isotropic, uniformly distributed, volume neutron source is created in the cell labeled "Plasma" in Figure 26.9. The energies of the source neutrons are assigned using the energy spectrum for deuterium–tritium fusion neutrons included in MCNP6.

In each of the 125 cells in each of the 10 nodes, tallies are placed for flux, absorption, fission, and $n,2n$ reactions. Direction-binned surface tallies are also placed on the outer surface of each node. The tallies considered to be of high relevance pass MCNP's ten statistical checks. For example, the absorption tallies in the sodium and the fission tallies in the fuel pass all of the tests. The resulting output file is quite large and contains results for over 5,000 separate tallies. A macro-enabled spreadsheet is used to automatically import the raw tally data from the MCNP output file. The spreadsheet condenses the data and calculates the effective kinetics parameters for each of the ten nodes.

To calculate $S_{fus,j}$, the same source definition as before is used (the isotropic, uniformly distributed, volume neutron source in the plasma cell), but the tallies are changed. A direction-binned surface tally is applied to the surface of each node, and the importance of each node is set to 0. Any plasma neutrons entering a node are tallied and immediately destroyed. This prevents the surface tally from double or triple counting any neutrons. Sixty million particles are run, and the tallies pass all statistical checks. This gives good-enough statistical accuracy. There are only ten tallies in the output file, so an automated spreadsheet is not required.

A bit more effort is required to calculate the coupling coefficients α_{kj}. It is a two-part process. Part one uses the first source described: the isotropic fusion neutron source. Instead of tallies, the surface source writer in MCNP6 is used. The surface source writer stores the position, direction, weight, and energy of every particle crossing a designated surface. The surface source writer is applied to the surface of node 1. Doing so enables the recording of the steady-state inward and outward neutron fluxes of node 1. Ten million particles are run. For part two, the recording is used as a surface source on the surface of node 1. The importances of all nodes are set to 0, and surface tallies are set on the surface of each node. This ensures the tallies do not count a neutron more than once. Symmetry ($\alpha_{kj} = \alpha_{jk}$) is assumed, so it is only necessary to calculate this for node 1. One hundred million particles are run. There are only ten tallies in the output file, so no automation is required.

26.7.3   Calculation of feedback effects

The MCNP model calculates how various perturbations affect the terms of the nodal kinetics equations. The perturbations considered are core grid plate expansion, fuel axial expansion, fuel bowing, thermal Doppler broadening, and sodium expansion and voiding. The core grid plate expansion, axial expansion, and fuel bowing perturbations are manifested as changes in geometry. The sodium expansion and voiding is manifested as sodium density changes in the nodes and pools. The Doppler broadening is manifested as different cross-section sets that have had their absorption resonances broadened at different temperatures. Node and pool temperature changes cause these perturbations to occur, which in turn affects the terms in the kinetics equations. Table 26.9 shows how the parameters Λ_{fj}, Λ_{2nj}, $S_{fus,j}$, $l_{e,j}$, and $l_{a,j}$ change with node temperature. It is assumed the parameters in Table 26.9 are functions only of node j's temperature and are not affected by the temperature in adjacent nodes. Several calculations showed that while node j is affected by temperature changes in adjacent nodes, the effect is at least one order of magnitude smaller than the effect of a temperature change in node j. Table 26.10 shows how the coupling coefficients change with pool temperature. Note that the fusion neutron source into each node increases with the temperature of the sodium in each node because the intervening sodium density decreases.

For any change in temperature, the MCNP model simultaneously incorporates the effects of all the perturbation types mentioned earlier. This is why each node is

Table 26.9   Kinetics coefficients at several node temperatures

Kinetics coefficient	Node j temperature			
	529 K	829 K	1,129 K	1,429 K
Λ_{fj} (s)	4.289E–07	4.299E–07	4.284E–07	4.272E–07
Λ_{2nj} (s)	3.672E–04	3.480E–04	3.478E–04	3.472E–04
$S_{fus,j}$ (cm^{-3}·s^{-1})	1.973E+11	2.009E+11	2.046E+11	2.083E+11
$l_{e,j}$ (s)	2.638E–07	2.598E–07	2.592E–07	2.586E–07
$l_{a,j}$ (s)	5.729E–07	5.757E–07	5.734E–07	5.717E–07

Table 26.10   Coupling coefficients at several pool temperatures

Coupling coefficient	Pool 6 temperature			
	529 K	829 K	1,129 K	1,429 K
$a_{6,1}$	0	0	0	0
$a_{6,2}$	0	0	0	0
$a_{6,3}$	0	0	0	0
$a_{6,4}$	0.00057	0.00059	0.00061	0.00064
$a_{6,5}$	0.02753	0.02804	0.02865	0.02936
$a_{6,6}$	0.74210	0.73755	0.73236	0.72640
$a_{6,7}$	0.02753	0.02804	0.02865	0.02936
$a_{6,8}$	0.00057	0.00059	0.00061	0.00064
$a_{6,9}$	0	0	0	0
$a_{6,10}$	0	0	0	0

subdivided into 125 cells and their materials homogenized. The subdivision and homogenization allow for changes in geometry and material properties to be modeled easily. The material density and composition of each of the 1,250 cells in the model are modified to reflect whatever change is desired. The MATLAB program UPM automates this process. UPM takes as input the temperature of each node, the temperature of each pool, and the physical displacements of each cell (i.e., geometry changes due to fuel bowing, grid plate expansion, or axial expansion). It calculates the new material densities and compositions for every cell in the model and writes this information to a text file in MCNP input format. It is easy to generate new MCNP input files for even the most complex combinations of perturbations. The following paragraphs describe how this MATLAB program accounts for each perturbation type.

UPM uses the core temperature to choose the cross section set to be used for that node's materials. Doppler-broadened cross-section sets for temperatures of 600, 829, 1,200, and 1,500 K have been created. They were produced using the NJOY [8] and the ENDF 7.0 [9] libraries. These temperatures are slightly different from the temperatures used in Tables 26.9 and 26.10, but the difference is small.

In SABR, sodium voiding refers to two different things: voiding in the core and voiding in the pool. The sodium in the core is likely to change temperature and density at a different rate than the sodium in the pool. Sodium voiding in the core affects the absorption and leakage terms of the nodal kinetics model. Sodium voiding in the pool affects the coupling and external source terms. UPM incorporates these changes by scaling the core and pool sodium densities separately based on their separate temperatures.

Before MCNP can determine how changes in geometry affect the kinetics terms, it needs to be understood how changes in temperature affect the geometry. Simple hand calculations are used to determine the physical displacements caused by grid plate expansion and axial expansion. It is assumed the core grid plate expands and contracts isotropically with changes in temperature. The change in temperature is multiplied by the coefficient of thermal expansion to find the strain. To find the x-dimension displacements of a cell, the strain is multiplied by a scale factor. The scale factor is the distance between the x-coordinate of the center of the grid plate and the x-coordinate of the center of the cell in question. This is done for every cell in the model. The process is repeated for the y-dimension as summarized in Figure 26.10.

For the case of axial expansion, the same method that was used with the grid plate expansion is employed. The strain is calculated and multiplied by a scale factor. Here, the scale factor is the distance from the bottom of the core to the center of the cell in question. This gives the z-dimension displacements.

UPM uses these displacements to simulate material relocation. It does this by looping over each cell in the model and adding/removing materials where appropriate. The actual grid geometry of the MCNP model never changes; only the material compositions and densities of the node cells change. UPM accounts for fuel bowing in the same way it accounts for grid plate expansion and axial expansion. It uses displacements to simulate material relocation. The difference

Core Grid Plate

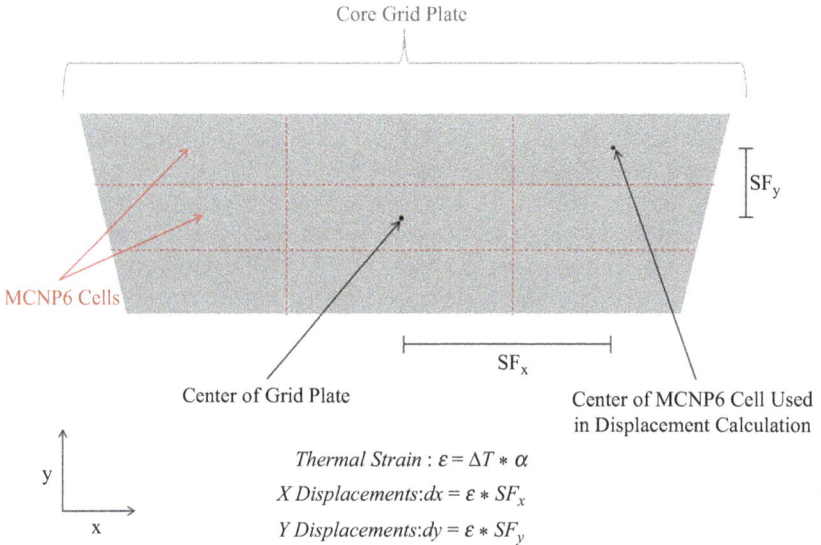

MCNP6 Cells

SF_y

SF_x

Center of Grid Plate

Center of MCNP6 Cell Used
in Displacement Calculation

y

x

Thermal Strain : $\varepsilon = \Delta T * \alpha$

X Displacements: $dx = \varepsilon * SF_x$

Y Displacements: $dy = \varepsilon * SF_y$

Figure 26.10   Diagram of displacement calculation for grid plate expansion

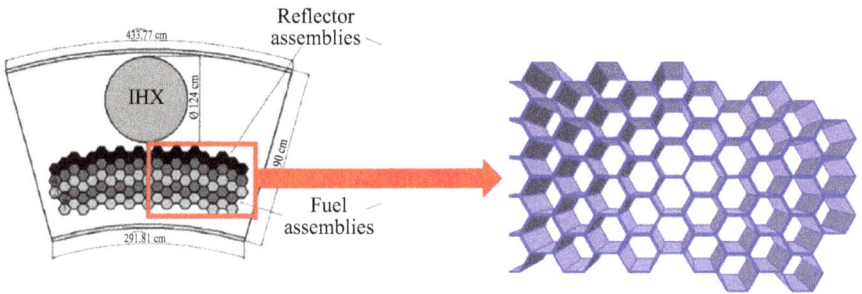

433.77 cm

IHX

Ø124 cm

Reflector
assemblies

90 cm

Fuel
assemblies

291.81 cm

Figure 26.11   Geometry of fuel bowing model

here is that the displacements are much harder to calculate. Fuel bowing is a highly non-linear process and requires a sophisticated model.

Fuel bowing is quantified in SABR using the Structural Mechanics Module in COMSOL Multiphysics. This is a separate COMSOL model that is not a part of the dynamics model. This model calculates the fuel assembly displacements in a single core at various power-to-flow ratios. The model uses a half-core geometry as shown in Figure 26.11 with a symmetry boundary condition to reduce computation time. The geometry contains all of the fuel assemblies and SiC flow channel inserts for a single core. Fuel pins are not included because they do not appreciably affect fuel bowing. The triangular swept mesh shown in Figure 26.12 is used on both the ducts and the inserts. The model accounts for geometric non-linearity, duct-to-duct contact, and duct-to-insert contact. A constrained boundary condition is applied to the bottom of the ducts. This is a simplifying assumption used for now because SABR's grid plate has not been designed yet. When more detail is available, the bottom boundary condition will need to be updated to a more realistic stiffness because the stiffness of the bottom boundary condition can have a significant impact on the fuel assembly displacements.

A 3D temperature distribution is assigned to the mesh using the MATLAB program comFBtemp. It calculates the temperature at every mesh point on the inner surfaces of the ducts and the SiC inserts by assuming it is the same temperature as the sodium adjacent to it. The program takes in the x-, y-, z-coordinates of each mesh point and calculates the temperature of sodium by integrating the amount of energy added to it as it travels up through the core in the z-direction. This requires the 3D power distribution of the core. The MCNP model of SABR can produce exactly that by using FMESH tallies. This has nothing to do with the nodal kinetics model; it is just an additional task the MCNP model can perform. The core is broken into 5,301 cells, and the fission reactions in each cell are tallied. A 3D power distribution is constructed using the tally results. A single power profile is calculated for the reference case. It is assumed the profile does not change during a transient.

The fuel bowing displacements are calculated for a power-to-flow ratio of 4. This is estimated as the highest ratio SABR can handle before design failure. The displacements for the z-dimension are not considered in the fuel bowing model because they are accounted for in the axial expansion perturbation model. The fuel

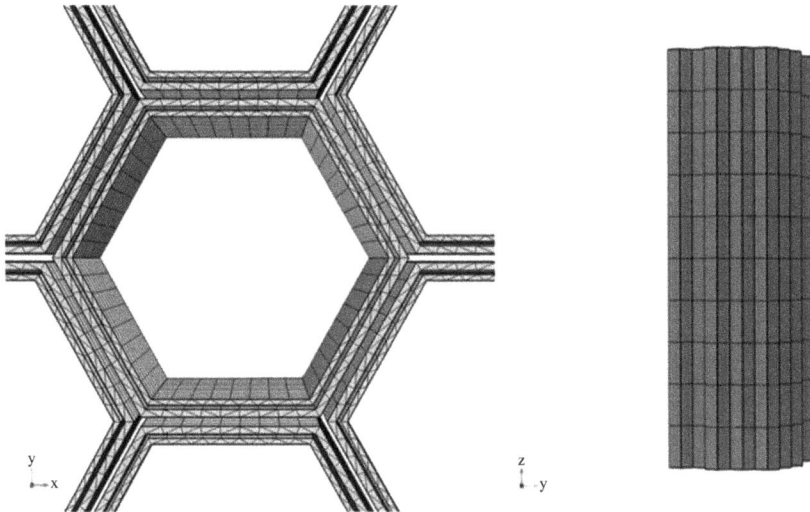

Figure 26.12 Top-down and side views of fuel bowing mesh

bowing for the most severe case possible is only a few millimeters. When these displacements are input to the MCNP model, no appreciable changes are seen in the kinetics parameters. Because of this, fuel bowing is not accounted for when performing the dynamic safety analysis of SABR.

26.7.4   Thermal-hydraulic model

The thermal-hydraulic component of the dynamics model utilizes COMSOL's Pipe Flow Module and Heat Transfer in Solids Module. Each of the ten pools is modeled separately. When modeling a pool, the entire primary loop (including the temperature distribution in the fuel, cladding, and primary heat exchanger) is considered, and only a small part of the secondary loop is considered. In the secondary loop, only the flow of sodium through the primary heat exchanger is modeled. Inlet and outlet boundary conditions are placed just outside the heat exchanger's secondary-side inlet and outlet. It is assumed the remaining power conversion and heat rejection cycles match those of a typical pool-type, sodium-cooled fast reactor design.

The Pipe Flow Module is used to create what is essentially a loop-type model of a SABR core. A special boundary condition in the loop accounts for additional heat capacity of the sodium pool. The structure of the loop-type model is similar to a RELAP5 model. It is a loop composed of sodium-filled pipes. Various boundary conditions simulate heat transfer from the core, across the heat exchanger, and into the sodium pool. A diagram of one of the ten loops in our model is shown in Figure 26.13.

26.7.5   Modeling the coupled cores

For each core, a single characteristic fuel pin is modeled. Two-dimensional (2D) axisymmetry is assumed as shown in Figure 26.14. The nodal kinetics model sets the power level in each of the ten different characteristic pins, and the Heat Transfer in Solids Module calculates the temperature distribution across the pins and cladding by solving the conduction equation in (26.4) for the case of 2D axisymmetry [10]:

$$\rho c_p \frac{\partial T}{\partial t} + \bar{u}_{\text{trans}} \cdot \nabla T + \nabla \cdot (q + q_r) = -\alpha T : \frac{d\bar{S}}{dt} + Q, \tag{26.4}$$

where

ρ = density
c_{p} = specific heat capacity at constant stress
T = absolute temperature
$\bar{u}_{\text{trans}}$ = velocity vector of translational motion
q = heat flux by conduction
q_r = heat flux by radiation
α = coefficient of thermal expansion
S = second Piola–Kirchhoff stress tensor [11]
Q = heat source.

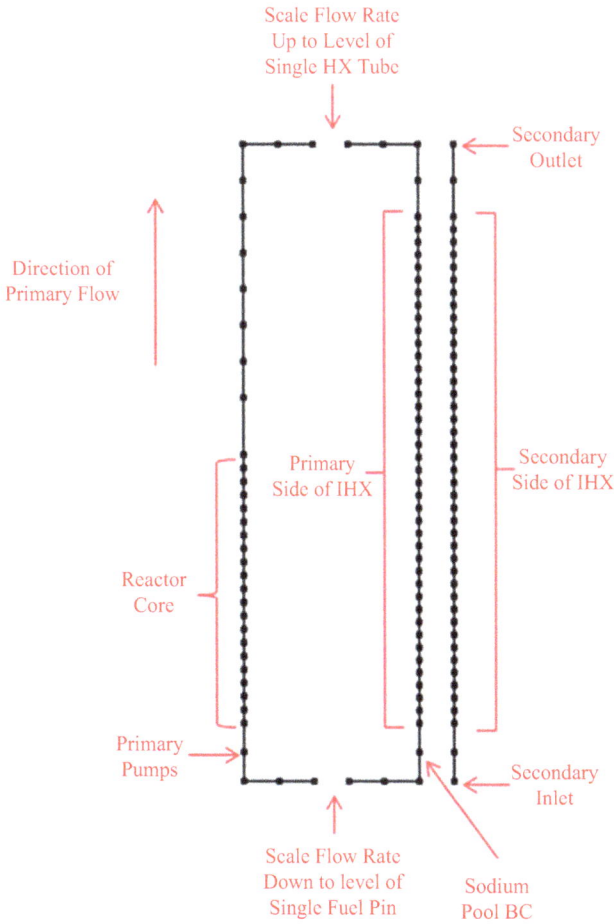

Figure 26.13   Diagram of sodium loop

For every point shown in Figure 26.14, the Pipe Flow Module calculates the heat transfer to the sodium coolant by solving the convection heat transfer equation:

$$q'' = hA(T_s - T_\infty) \qquad (26.5)$$

where

q'' = heat flux
h = heat transfer coefficient
A = surface area
T_s = surface temperature.
ρ = density
$\bar{u}$ = velocity vector
p = pressure

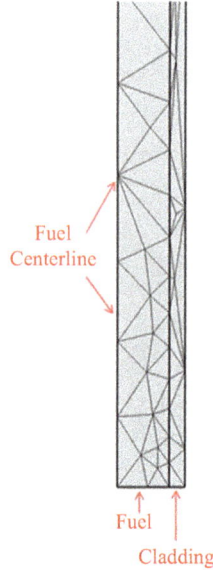

Figure 26.14   Geometry of fuel pin

$\bar{\tau}$ = viscous stress tensor
F = volume force vector
c_p = specific heat capacity at constant pressure
T = absolute temperature
q = heat flux vector
Q = heat source
$\bar{\bar{S}}$ = strain-rate tensor.

This approach requires various empirical correlations that are described in the later section.

26.7.6   Modeling the heat exchanger

The same procedures and equations described above are used to model the heat transfer in the heat exchanger. A single heat exchanger tube is modeled with the assumption of 2D axisymmetry as shown in Figure 26.15. The Pipe Flow Module calculates the surface temperatures on both sides of the tube. The Heat Transfer in Solids Module uses the surface temperatures to calculate the temperature distribution and heat transfer across the tube. This gives the inlet and outlet temperatures for both sides of the heat exchanger.

In each pool, there are 5,700 tubes in the heat exchanger and 37,520 fuel pins in the core. One of each is being modeled, so the mass flow rates are scaled proportionally to accurately model the system. For the core side of the loop, the flow rate is scaled to a value corresponding to single fuel pin. This is 1/37,520th of the core's primary mass flow rate. For the heat exchanger side of the loop, the flow rate is scaled to a value

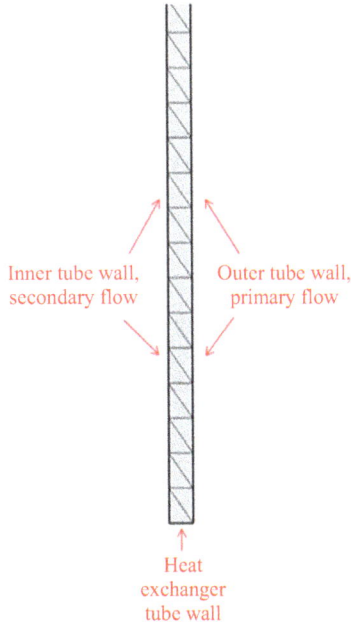

Figure 26.15   Heat exchanger tube geometry

corresponding to a single heat exchanger tube. This is 1/5,700th of the core's primary flow rate.

26.7.7   Modeling the sodium pools

The additional heat capacity of the sodium pool is accounted for by using the MATLAB program PoolTemp. As sodium in the primary loop exits the heat exchanger, the Pipe Flow Module calls PoolTemp to bring the sodium to thermodynamic equilibrium with the current pool temperature. This new temperature is set as the inlet temperature for the core. This is a three-step process.

Step one increases the mass and energy of the pool as sodium enters it from the primary outlet of the heat exchanger:

$$m_{\text{pool},2} = m_{\text{pool},1} + \dot{m}_{\text{in}}\Delta t$$

and

$$E_{\text{pool},2} = E_{\text{pool},1} + c_{p,\text{in}}T_{\text{in}}\dot{m}_{\text{in}}\Delta t$$

where

$m_{\text{pool},1}$ = mass of the pool at the beginning of the time step
$m_{\text{pool},2}$ = mass of the pool in the middle of the time step
$E_{\text{pool},1}$ = internal energy of the pool at the beginning of the time step
$E_{\text{pool},2}$ = internal energy of the pool in the middle of the time step

$\dot{m}_{\text{in}}$ = mass flow rate of incoming sodium
ΔT = size of the time step
$c_{\text{p,in}}$ = specific heat capacity evaluated at T_{in}.

The second step assumes thermodynamic equilibrium in the pool and calculates its average temperature using the temperature of sodium exiting the pool and entering the core:

$$T_{\text{out}} = \frac{E_{\text{pool,2}}}{m_{\text{pool,2}} c_{\text{p}}}$$

where c_{p} is the specific heat capacity evaluated at T_{in} and T_{out} is the temperature of outgoing sodium.

The third step decreases the mass and energy of the pool as sodium leaves it to enter the core inlet:

$$m_{\text{pool,3}} = m_{\text{pool,2}} - \dot{m}_{\text{out}} \nabla t$$

where

$m_{\text{pool,3}}$ = mass of the pool at the end of the time step
$E_{\text{pool,3}}$ = internal energy of the pool at the end of the time step
$c_{\text{p,out}}$ = specific heat capacity evaluated at T_{out}.

26.7.8   Thermal property data and empirical correlations

COMSOL is given the material properties for the fuel, cladding, and sodium as well as the empirical correlations for the friction factor and Nusselt number. It is assumed the properties of the fuel, cladding, and heat exchanger tube are constant. The same data used for the heat exchanger tube are used for the cladding. The properties are shown in Table 26.11 [12,13].

The ANL sodium property correlations are used, and they are in (26.4)–(26.9).

26.7.9   Calculation of nodal heat transfer terms

The kinetics model requires both node and pool temperatures as input. A node is defined as the fission core in each pool. This includes all fuel and reflector assemblies but not the heat exchanger and the surrounding sodium in the pool. The average node temperature is calculated using

$$T_{\text{node}} = \frac{m_{\text{fuel}} T_{\text{fuel}} + m_{\text{clad}} T_{\text{clad}} + m_{\text{reflector}} T_{\text{reflector}} + m_{\text{sodium}} T_{\text{sodium}}}{m_{\text{fuel}} + m_{\text{clad}} + m_{\text{reflector}} + m_{\text{sodium}}} \qquad (26.6)$$

Table 26.11   Fuel and clad property data

	TRU fuel	Oxide dispersion-strengthened cladding
Thermal conductivity (k)	10 W/mK	30 W/mK
Density (ρ)	3,861 kg/m^3	7,692 kg/m^3
Specific heat capacity (c_{p})	738.15 J/kg·K	650 J/kg·K

where T and m are the temperatures and masses of the fuel, cladding, sodium, and reflector in one of SABR's cores. COMSOL calculates T_{fuel}, T_{clad}, and T_{sodium} by finding the average temperatures in the fuel, clad, and sodium domains. It is assumed that $T_{reflector}$ is the same temperature as the inlet temperature of the sodium because there is negligible heating in the reflector region. The masses of each component are hand-calculated constants that are hard coded into the COMSOL model.

The pool temperature is calculated. Thermodynamic equilibrium is assumed in the pool indicating that $T_{pool} = T_{out}$.

26.8 Verification tests

SABR's dynamics model was subjected to several verification tests. No experimental data for SABR exist, of course, so validation tests were unable to be performed. The verification tests were performed on individual subcomponents of the dynamics model: the neutron kinetics solver, thermal-hydraulic solver, and fuel bowing model. The dynamics model itself cannot be verified because there is no other model to compare it with. It is assumed that verification of the individual subcomponents of the dynamics model will suffice.

26.8.1 Neutron kinetics model

The MATLAB program NodalSolve is used to solve the nodal neutron kinetics equations over any desired time step. The nodal equations represent a stiff system because they are governed by multiple time constants that are different orders of magnitude: prompt neutrons, which change on the order of fractions of microseconds, and delayed neutrons, which change on the order of seconds.

The accuracy of the numerical solution was verified by taking a Laplace transform of the system and comparing it to the numerical solution. The system is too complicated to be easily solved with Laplace transforms, so it was simplified, and Mathematica was used to apply the transformations [17]. These tests were performed on single-node systems with a single delayed neutron group whose decay constant was varied from case to case. For brevity, only a single case is shown (Figure 26.16).

The governing equations for this case are

$$\frac{dn}{dt} = \left(\frac{1}{\Lambda_f} + \frac{1}{\Lambda_{2n}} + \frac{\alpha_{11}}{l_e} - \frac{1}{l_a} - \frac{1}{l_e} \right) n(t) + S_{fus} + \lambda c(t) \tag{26.7}$$

and

$$\frac{dc}{dt} = \frac{1}{l_e} n(t) - \lambda c(t) \tag{26.8}$$

The source term is reduced by 50% at $t = 0$, and the resulting Laplace solution is given by:

$$n(t) = 237162 + 233232e^{-431161t} + 66245.1e^{-2.79093t} \tag{26.9}$$

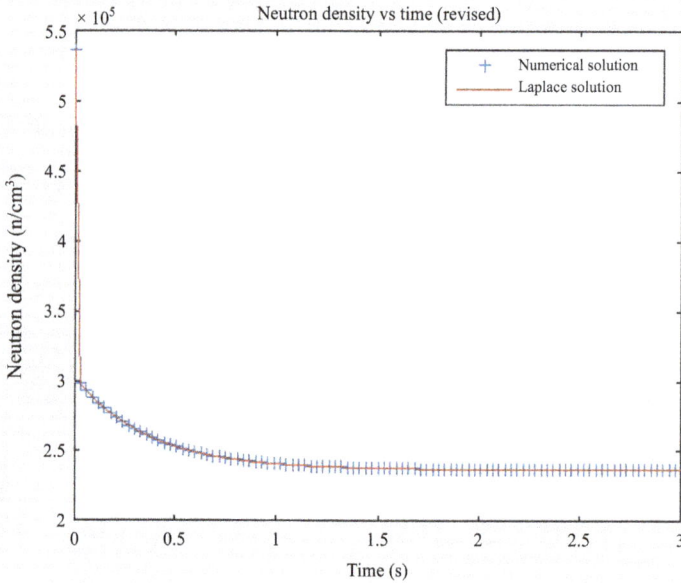

Figure 26.16   Test problem neutron density versus time (one-group delayed neutrons, 50% reduction in source strength at t = 0)

Figure 26.16 compares the two solutions, which agree quite well, providing verification for the neutron kinetics solver.

26.8.2   Thermal-hydraulic model

RELAP5 (see [18] for an introduction of RELAP5) was used to test the thermal-hydraulic component of our model. Because RELAP5 cannot handle the complexity of the SABR model, it was simplified to a single-node, loop-type core with no feedback. The remaining core geometry, composition, and thermal-hydraulic conditions are identical to the SABR design. The steady-state temperatures and pressure drops were compared, and the transient maximum fuel temperatures and maximum coolant temperatures were compared for every type of accident scenario in the dynamic safety analysis.

The steady-state temperature comparisons are shown in Table 26.12. The temperatures and pressure drops agree well with RELAP5.

For the sake of brevity, this paper shows only one of the six transient cases compared. The maximum fuel and coolant temperatures for a 50% LOFA in core 1 are shown in Figure 26.3.

The transient results agree well, and the thermal-hydraulic model is considered verified.

26.8.3   Fuel bowing model

COMSOL was used to model the structural mechanics component of the fuel bowing analysis. The fuel bowing model was verified by using it to model an

Table 26.12 Steady-state parameters

	RELAP5	COMSOL
Average fuel temperature (K)	725	720
Maximum fuel temperature (K)	801	802
Core inlet temperature (K)	619	619
Core outlet temperature (K)	759	759
Secondary inlet temperature (K)	590	590
Secondary outlet temperature (K)	739	739
Core pressure drop (kPa)	70	71
Primary pressure drop (kPa)	264	266
Secondary pressure drop (kPa)	33	33

Table 26.13 Comparison of displacement estimations

	COMSOL model	IAEA problem	Percent difference
x displacement	−11.66 mm	−12.36 mm	5.7%
z displacement	32.91 mm	31.86 mm	3.2%

International Atomic Energy Agency (IAEA) structural mechanics verification problem [19]. The verification model used similar settings for the mesh and physics solvers as in SABR's fuel bowing models.

The problem solution is a hand calculation made by the IAEA's International Working Group on Fast Reactors. The problem contains a single-fuel assembly made of stainless steel, and it uses Cartesian coordinates with the origin centered on the bottom face of the duct. The z-axis is parallel with the duct axis. The duct wall is 3 mm thick with an across-flats dimension of 132.9 mm. The material properties are $T_{ref} = 20°C$, $\alpha = 18.6 \times 10^{-6}/°C$, $\nu = 0.3$, and $E = 170$ GPa. The temperature field is described by the piecewise function below with $a_0 = 212.5°C$, $a_1 = 0.500°C/mm$, $a_2 = 0.125°C/mm$, and $a_3 = -3.333 \times 10^{-4}°C/mm$:

$$\begin{cases} 400°C & z \leq 1,500 \text{ mm} \\ a_0 + a_1 x + a_2 z + a_3 zx & 1,500 \text{ mm} < z < 2,500 \text{ mm} \\ 525°C - \dfrac{50°C}{150 \text{ mm}} & 2,500 \text{ mm} \leq z \leq 4,000 \text{ mm} \end{cases} \quad (26.10)$$

A similar swept, triangular mesh is applied to the model. The same minimum/maximum parameters as the SABR fuel bowing model are used, but the axial mesh is refined a bit. The axial temperature profile in the problem varies more strongly than the axial temperature profile of SABR. The same physics solvers are used as in the SABR model, and geometric non-linearity is included. The x and z displacements at the top of the assembly are calculated. The results are shown in Table 26.13.

The results agree well, and the fuel bowing model of SABR is considered to be verified.

26.9   Dynamic safety analysis of SABR#2

The temperatures shown in Table 26.14 are used as the initial conditions for temperature. A maximum time step of 0.1 s was used for every accident scenario with a total time of 125 s unless more time was needed; 125 s is the minimum time required for the coolant flow rates to coast down and reach a constant value. SABR uses electromagnetic sodium pumps whose power source is connected to a mechanical flywheel system. When power is lost, the pumps have a 10-s halving time. The coastdown profiles for the 50% and 100% accidents are shown in Figures 26.18 and 26.19. The COMSOL dynamics model uses these data to control the primary and secondary flow rates as a function of time.

When both pumps fail, the flow rate does not coast down all the way to zero. There is some natural circulation in the core. The method prescribed by Todreas and Kazimi was used to estimate natural circulation in the primary loop [20]:

$$\dot{m} = \left(\frac{2\beta \dot{Q}_H g \Delta L \rho_0^2}{c_p R}\right)^{\frac{1}{3-n}}$$ (26.11)

where

β = coefficient of thermal expansion defined by $\partial \rho / \partial T \times (1/\rho)$

$\dot{Q}_H$ = core power

g = acceleration due to gravity
ΔL = height difference between the center of the core and the center of the heat exchanger
ρ_0 = reference density of sodium
c_p = average heat capacity of sodium
n = 0.2 for turbulent flow
R = hydraulic resistance coefficient.

Table 26.14   *Fusion neutron source strength increase accident summary*

	BOL	BOC	EOC
Steady-state fusion power level (MW)	73	242	370
Fusion power leading to coolant boiling (MW)	121.2	398.1	563.9
Maximum tolerable increase in fusion power (MW)	48	156	194
Maximum fuel temperature, K	1,265	1,336	1,337

Note: Coolant boiling occurs at 1,156 K and fuel melting occurs at 1,473 K.

This natural convection model assumes steady-state conditions. While the safety analysis is not simulating steady-state conditions, this model can still be used. Any accident results in either failure or new steady-state conditions. If the accident results in new steady-state conditions, the use of the model is justified. If the accident results in failure, the use of the model is irrelevant. This dynamic safety analysis is only interested in determining if and when failure occurs, not what happens after failure. The model predicts that natural circulation through the primary loop will be 2% and 0.5% of the normal flow rate for full power and decay power levels, respectively. The decay power for SABR's current design has not been explicitly calculated. It is conservatively assumed that the decay power is a constant 7% of full power. This assumption is based on studies of decay power in TRU fuel [21].

Each case in the dynamic safety analysis was evaluated on a pass/fail basis. A case failed if the coolant reached its boiling point of 1,156 K or the fuel reached its melting point of 1,473 K. Unless otherwise noted, all accident-initiating events were run in core 1 only. Cores 2 through 10 remained unaffected.

26.9.1 Accident scenarios and corrective actions

The accidents we simulated were LOFAs, LOHSAs, and LOPAs. The primary and secondary loops of each SABR pool have two pumps, so a 50% LOFA/LOHSA simulates the loss of one pump in the primary/secondary loop while the 100% LOFA simulates the loss of both pumps in the primary loop. In core 1, 50% LOFAs/LOHSAs and 100% LOFAs/LOHSAs/LOPAs were run with no corrective action taken. These same cases were also run with two different corrective actions: fusion plasma shutdown and control rod insertion.

SABR's plasma is maintained by external power sources that heat the plasma and drive confining current in the plasma. Simply switching off these external power sources would reduce the neutron source strength as the plasma cooled down. Since the energy confinement time in the plasma is on the order of seconds, the neutron source strength might be expected to decay exponentially with a time constant on the order of seconds. There are constraints on ramping down magnetic fields that might lead to the necessity of slower neutron source decays. Initial results suggest the plasma power can be reduced to 0 within 10 s. For the purpose of the present work, the conservative assumption of an exponential decay with a time constant of 10 s is made.

Most fast reactor control systems initiate shutdown when certain parameters such as core temperature or power-to-flow ratio reach a certain value. Once the threshold is hit, the control system takes on average 2 s to respond. For SABR, our threshold is pump failure. When a pump fails, it is assumed the reactor control system takes 2 s to notice and another 3 s to respond by initiating a plasma shutdown. Thus, the plasma shutdown is initiated 5 s after a pump in either the primary or secondary systems in any of the fission cores stops.

When using control rods as a corrective mechanism, the plasma power remains at 100%, and the control rods are inserted to decrease the fission power several seconds after an accident begins. It is assumed it takes 2 s for the reactor control system to notice the pump failure and 1 s to initiate a control rod drop in all ten fission cores. It is assumed the control rods take 2 s to fully insert once they are dropped. These values are approximate to normal fast reactor control systems.

To date, SABR's design does not explicitly include control rods, so a rod bank similar to the one in ANL's Advanced Burner Test Reactor [22] (ABTR) is used. The ABTR uses a bank of ten control rods that provide up to −40.19 $ of reactivity. The ABTR is a good analog because it has a power, fuel composition, and delayed neutron fraction comparable to SABR. A SABR core generates 300 MW(thermal) of power while the ABTR generates 250 MW(thermal). SABR's fuel is TRU while the ABTR's fuel is a mixture of TRU and uranium. SABR's delayed neutron fraction is $\beta_{\text{eff}} = 0.0030$ and ABTR's is $\beta_{\text{eff}} = 0.0033$.

If the rod distribution shown in Figure 26.17 is used, nine control rods can be fit into each SABR core. Each of the ten control rods in the ABTR have different reactivity worths, but here the average is taken, and it is assumed each rod has the same worth. Under that assumption, each rod would be worth 4.02 $. Using $ *of reactivity* $= \Delta\rho/\beta_{\text{eff}}$ to convert that to a reactivity per rod yields $\Delta\rho = -0.0134$ per rod.

SABR's total control rod worth is estimated by dividing $\Delta\rho$ by SABR's β_{eff} and multiplying by nine rods. This yields a control rod worth of −39.67 $ per pool lowering k_{eff} from 0.92185 to 0.83047, which corresponds to a 15.985% decrease in the absorption lifetime l_{a}.

26.9.2   Accident results

Figure 26.18a and 26.18b show that SABR is passively safe (i.e., avoids failure without any external corrective action taken) against 50% LOFAs. No fuel melting or coolant boiling occurs, and the temperatures remain well below the threshold of failure. A similar pair of curves (not shown) demonstrate that SABR is passively safe against 50% LOHSAs as well.

Figure 26.18a shows that feedback effects would have a negligible effect on reducing the core power level in the 100% LOFA, that control rod insertion would reduce the power level to 36% of full power, and that shutting off the plasma neutron source would reduce the core power level to the decay heat level

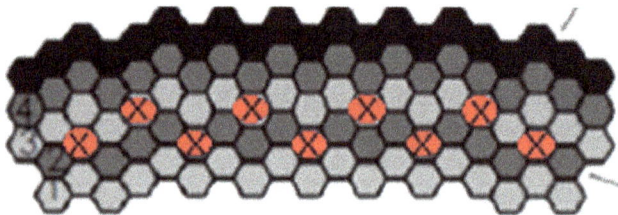

Figure 26.17   Control rod distribution (x)

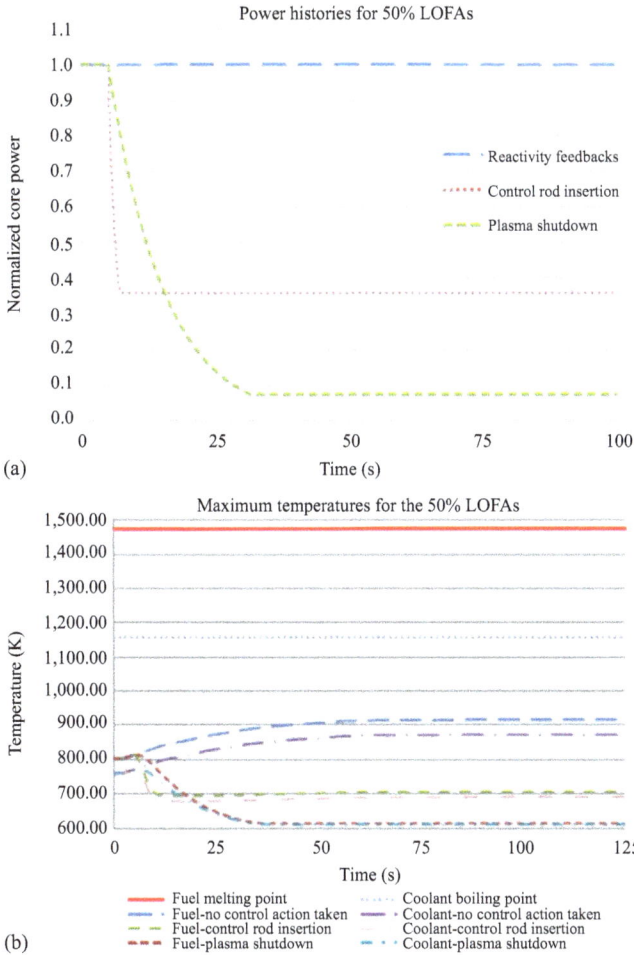

(a)

(b)

Figure 26.18 (a) Power histories for 50% LOFAs and (b) maximum temperatures for the 50% LOFAs

of 7% of full power. Similar power profiles were obtained for the 100% LOHSA and LOPA, with the exception being that the reactivity feedbacks reduced the core power level slightly more in the LOHSA case than with the 100% LOFA and 100% LOPA.

Figure 26.19b shows that SABR reaches the point of failure during a 100% LOFA with both coolant boiling and fuel melting regardless of any corrective action that is applied. When no corrective action is applied, reactivity feedbacks only slightly reduce the power level, and the core reaches the point of failure after 27 s. When control rods are inserted, they reduce the core power to 36% of full power. The reactor continues to operate at this reduced power level until it reaches the point of failure after 59 s. When the plasma is shut down, the plasma power drops to zero, and the reactor power level drops to the decay heat level of 7% of full power after 33 s, but the decay power is not adequately removed by natural circulation. The remaining decay power causes the coolant to boil and the fuel to melt at 208 and 275 s, respectively, after accident initiation. These matters need to be dealt with in future iterations of the SABR, or other FFH designs. *Again, putting different pumps on different electrical circuits will help minimize the possibility of such large % LOFAs, and should be emphasized in future design activities.*

Figure 26.20 shows that SABR fails during the 100% LOHSAs even with corrective actions. The times to failure are longer than those for the 100%

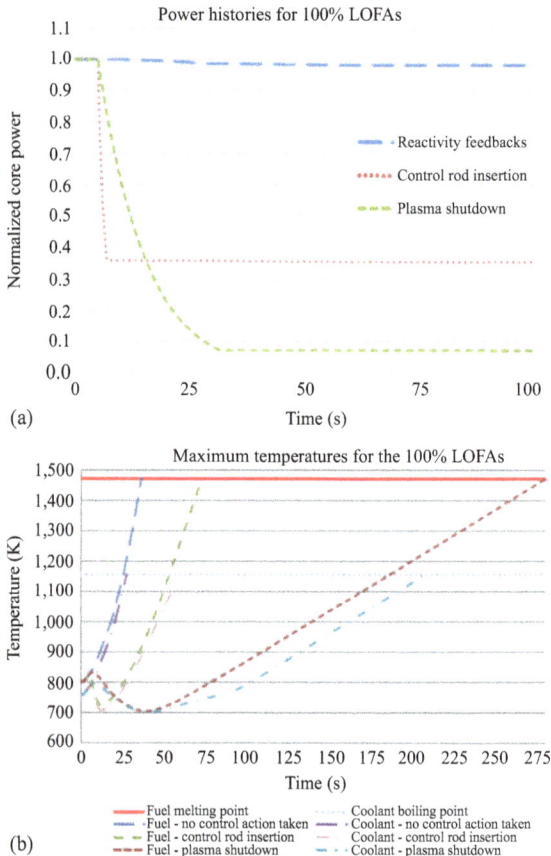

Figure 26.19   (a) Power histories for the 100% LOFAs and (b) maximum temperatures for the 100% LOFAs

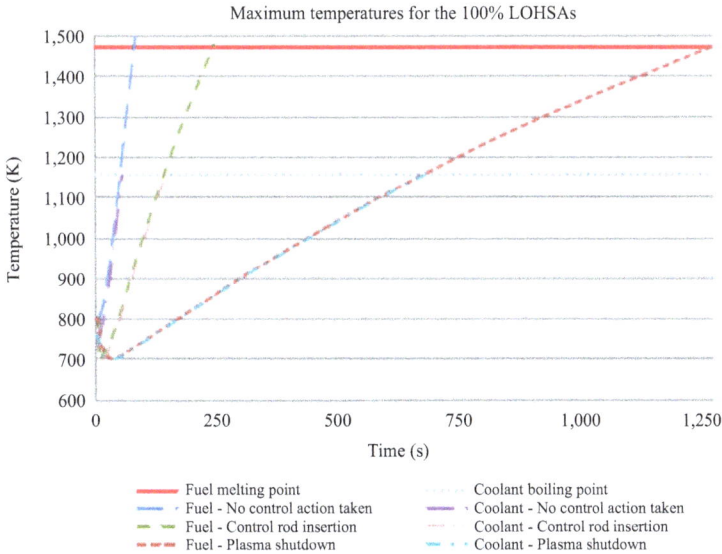

Figure 26.20 Maximum temperatures for the 100% LOHSAs

LOFAs because of the thermal capacitance of the pool. The more sodium there is in the pool, the longer is the time to failure. The reactivity feedbacks are slightly more impactful than for the 100% LOFAs, but they are still quite weak. The control rods and plasma shutdown mechanisms perform similarly as they did during the 100% LOFAs. The control rods do not provide enough negative reactivity to shut down the core. The plasma shutdown successfully shuts down the fission core and lowers the power level to decay heat levels before failure; however, the residual decay heat is not removed, and the reactor reaches the point of failure in about 30 min. This can be corrected (and will be in the next design iteration).

Figure 26.21 shows that SABR fails for the 100% LOPAs with or without corrective actions. Here, a worst case is taken for the LOPA, namely, that the fission reactor and fusion neutron source have separate power supplies and only the fission reactor loses power. The results are almost identical to those of the 100% LOFAs because the reduction in primary flow effectively decouples the core from the heat exchanger and secondary cycle.

26.9.3 Neutronic coupling

As expected, the neutronic coupling among the cores is strong. The top plot in Figure 26.22 illustrates the effect of control rod insertion in core 5 on the power levels in each of the ten cores. Also shown for comparison are the core powers when control rods are inserted in all ten cores.

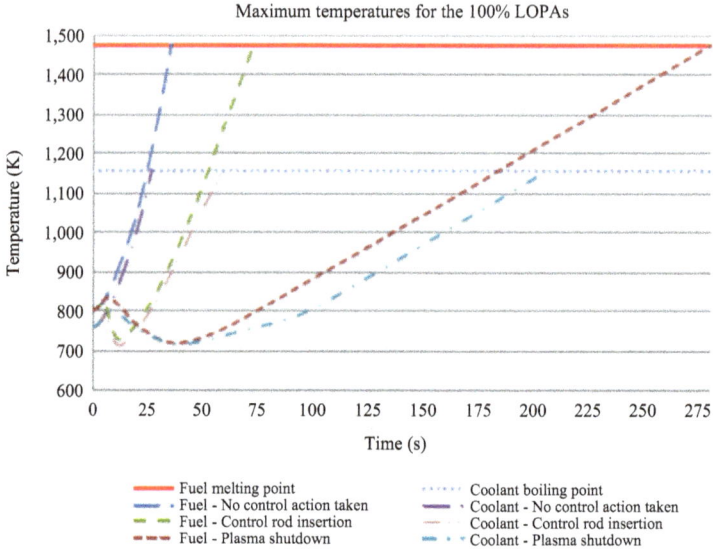

Figure 26.21   Maximum temperatures for the 100% LOHSAs

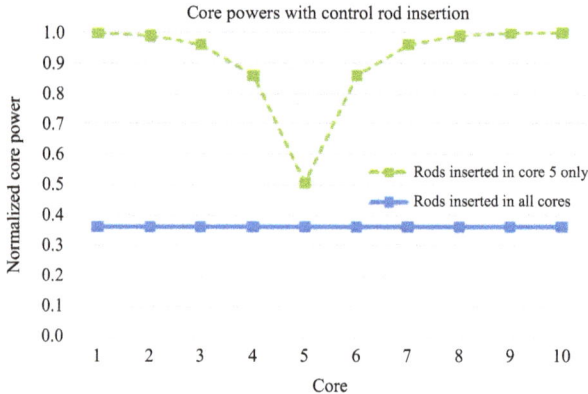

Figure 26.22   Comparison of power levels with control rod insertion

26.10   Discussion

A dynamic safety model has been created for SABR#2 using COMSOL Multiphysics for the thermal hydraulics, a nodal neutron kinetics model with parameters evaluated using Monte Carlo, and feedback models accounting for the effect of temperature changes on cross sections and on fuel movement. The model was verified by comparison against RELAP5, analytical solutions, and benchmark problems. SABR has been found to be passively safe (no coolant boiling or fuel melting even without control action) in accident scenarios that model the failure of 50% of the coolant pumps in either the primary or secondary loops. For the failure of 100% of the pumps in either the primary or secondary loops, the power could be quickly reduced to the 7% decay power level by shutting off power to the plasma neutron source, but the low level of natural circulation was insufficient to remove the accumulating decay heat, and coolant boiling followed by fuel melting eventually took place. *Future design work should address the decay heat removal in SABR.*

Control rods were found to be a less effective shutdown mechanism than switching off the power to the plasma source, and reactivity feedbacks were found to be a relatively ineffective shutdown mechanism in a subcritical system.

Several years prior to this work, a dynamic safety analysis was run on an older version of SABR's design [45]. RELAP5 was used to run LOFAs, LOHSAs, and LOPAs. A point kinetics model was used, and the reactivity feedbacks accounted for were Doppler broadening and sodium voiding. The results were similar to those found in this work, namely, that SABR could survive 25% pump failure accidents without control action, could survive 50% pump failure accidents in the secondary system (but not in the primary system) without control action, and could survive a LOPA because the neutron source also was turned off in a LOPA. The times to failure were marginally shorter than in the present work.

The results of this previous work were used to update SABR's design in 2014. There were two objectives of the redesign: (1) to improve SABR's safety performance by changing it from a loop-type core to a pool-type core and (2) to increase its practicality by more realistically integrating the fission and fusion reactor components together. Accomplishing the second objective involved changing the once annular and continuous fission core into ten separate core modules. Doing so totally changed the neutron kinetics of SABR. Originally, it could be assumed that any perturbation would affect the core uniformly, and point kinetics were used. With SABR's fission core now composed of ten physically separate but neutronically coupled cores, that assumption no longer applies. Conditions among the cores could conceivably vary and give rise to standing power oscillations or power tilts, if there was a driving mechanism for such. Thus, the nodal neutron kinetics model presented in this paper was developed to replace the point kinetics model.

In summary, this work differs from the previous work in three major ways: (1) nodal kinetics were used to model a ten-core reactor instead of point kinetics being used to model a single-core reactor; (2) a pool-type reactor is modeled instead of a loop-type reactor; (3) more reactivity feedbacks are included in the kinetics model. The present work accounts for Doppler broadening, sodium voiding, fuel and reflector axial expansion, core grid plate expansion, and fuel bowing, while the previous work only accounted for Doppler broadening and sodium expansion. It is also useful to compare SABR's performance with that of the Experimental Breeder Reactor-II (EBR-II). The choice of metal fuel and Na-pool design for SABR was influenced by the passive safety demonstration tests performed in EBR-II in 1986. In those tests, EBR-II exhibited full passive safety during uncontrolled 100% LOFAs and LOHSAs. These EBR-II results are different from those calculated for

SABR for two reasons: (1) SABR's neutron source and (2) SABR's lack of adequate decay heat removal.

SABR's operation with an external neutron source weakens the effect of reactivity feedbacks. In EBR-II, feedbacks generated negative reactivity that significantly affected the neutron balance. In SABR, the fission chain reaction is maintained by an external neutron source that offsets $\Delta k_{sub} = 1 - k \geq 0.03$ of reactivity, so the much smaller reactivity effects of feedback are relatively less important than they are in EBR-II.

Furthermore, once reactivity feedbacks shut down EBR-II, it was able to remove all decay heat through passive means. EBR-II incorporated additional heat exchangers whose sole purpose was to remove decay heat using natural circulation. SABR's current design has no explicit decay heat removal system, and that is the reason for fuel melting and coolant boiling even after SABR is successfully shut down by turning off the plasma. *Additional decay heat removal should be included in the next iteration of the SABR design.*

It was found that (a) the core power can be reduced to decay heat levels in a couple of seconds by turning off the neutron source heating power when any accident condition is detected; (b) a LOPA thus reduces the core to the decay heat level in a couple of seconds and natural circulation prevents core damage; (c) undetected LOFAs in which 50% of the primary coolant pumps fail can be survived without core damage, and only when 75% of the pumps fail does fuel melting occur at 8.4 s; (d) an undetected LOHSA with 50% loss of sodium flow in the intermediate loop can be survived without core damage, and only with 75% loss-of-sodium flow in the intermediate loop does fuel melting occur at 150 s; and (e) neutron source excursions due to inadvertent increases in plasma heating or fueling could be limited by operation near inherent plasma density and beta limits.

SABR can withstand the failure of up to one-quarter of the coolant pumps in either the primary or intermediate coolant loops without coolant boiling. If any additional coolant pumps fail, the coolant will either exceed or come very close to its boiling temperature. However, a failure of three-quarters of the coolant pumps in either the primary or intermediate coolant loops will result in fuel melting. With only 10 s before fuel melting begins, there is not enough time for operator intervention to terminate the transient, and this again indicates the need for an automatic control system and different pumps on different electrical systems.

Because SABR cannot sustain multiple pump failures without experiencing coolant boiling and possibly fuel melting, *coolant pumps in both the primary and intermediate coolant loops should be kept on entirely separate electrical systems from any other pump to ensure that a failure of one pump is not likely to be followed by a second pump failure.* After a single pump failure is detected, the neutron source should be quickly shut off. Even if all of SABR's pumps fail, if the neutron source is shut down at the same time, natural circulation will provide enough coolant mass flow to remove the decay heat being generated in the fission core.

The other category of transients is those affecting SABR's neutron population in the fission core. To prevent accidents related to inadvertent control rod removals from possibly causing coolant boiling in the fission core, two design changes are possible. The best option is to decrease the total control rod worth. Because control rods are used only for small changes in the fission power level and they are not essential to shut down the reactor, a decrease in control rod worth would not be a reduction in the SABR's level of safety.

Rather large increases in fusion neutron source strength can be tolerated before coolant boiling or fuel melting occurs. Throughout the fuel cycle, it would take an increase of at least 50% of the neutron source strength before SABR's heat removal system would be incapable of regulating core temperatures. In the case of EOC operation, the neutron source would have to exceed its maximum design strength before core failure occurs.

While some of the accidents simulated can cause damage in SABR if uncontrolled, core damage can be prevented for all transients by shutting off the neutron source. The fission core power level will quickly decrease to decay heat levels, leaving at most 7.1% power and enough coolant mass flow from natural circulation to cool the reactor until whatever caused the accident can be corrected. Whether a pump or a heat exchanger fails or a reactor operator utilizes the control rods improperly, simply shutting off the source neutrons will allow for a safe progression of the ensuing transient. *This would appear to be a major advantage of subcritical operation.*

Chapter 27

Space-dependent dynamics calculation model for a sodium pool-cooled fast transmutation reactor (SABR#2)

The subcritical advanced burner reactor #2 (SABR#2; Figure 27.1) reactor is a Na-cooled, pool-type, metal-fueled fast transmutation reactor, described previously.

In order to impact climate change over the next century, a massive source of clean, carbon-free energy is needed. In the authors' opinion, nuclear power provides the only technically and environmentally credible option. A major technical problem preventing the widespread expansion of nuclear power is spent nuclear

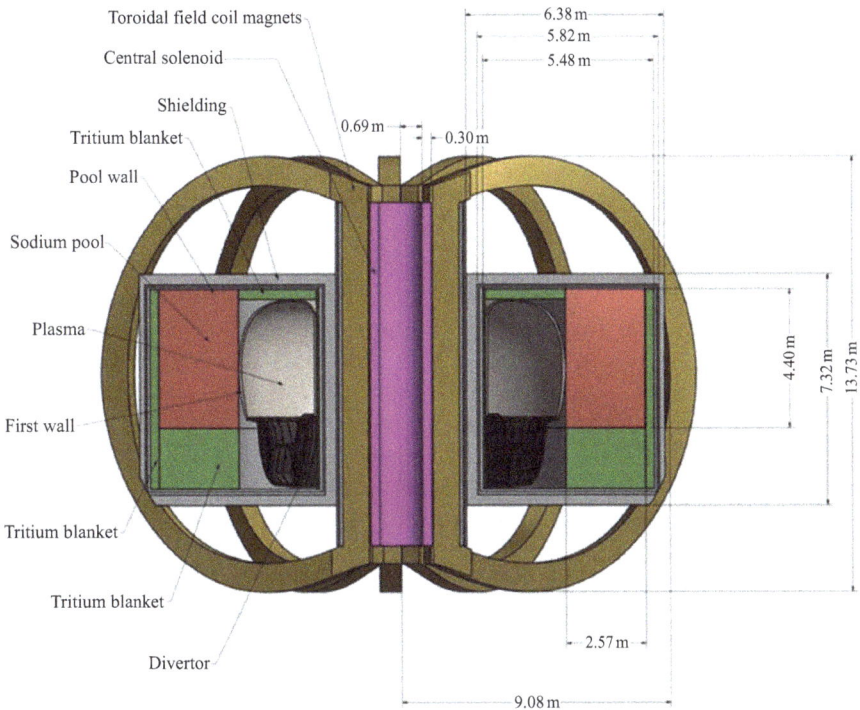

Figure 27.1 Overview of SABR#2

fuel (SNF). The US government currently has no plan to deal with SNF generated by commercial nuclear power plants. Plant operators have been forced to stockpile SNF on-site. The common suggestion to deal with SNF is the high-level waste repository (HLWR). A repository would have to store the SNF on the order of a million years until it is no longer significantly radioactive. Various organizations have proposed several HLWRs, but none of them are close to becoming a reality in the United States.

There is a more efficient way to dispose of SNF. The long-lived radioactive transuranics (TRU) that are intended to be contained within SNF can be destroyed by neutron fission in a fast burner reactor. The high-energy neutron spectrum of a fast reactor enables it to fission TRU and thereby transmute them into shorter-lived fission products. This process destroys TRU and produces more recoverable energy from the uranium fuel at the same time.

Critical fast burner reactors, which have a reactivity margin to a prompt critical runaway of $\Delta k_{crit} \approx \beta$, require some uranium ($\beta \approx 0.006$) driver fuel to be mixed in with the TRU ($\beta \approx 0.002$) in order to increase the reactivity safety margin to prompt critical $\Delta k \approx \beta$ for critical reactors. The requirement for uranium fuel reduces the amount of TRU burned in the reactor, and it also produces some TRU. This reduces the efficiency of critical fast burner reactors, relative to subcritical reactors for which $\Delta k_{crit} = \Delta k_{sub}$. On the other hand, a more efficient procedure is to have a reactor completely fueled with TRU operated subcritical with an adjustable external neutron source used to provide the extra neutrons to fission the TRU, to help maintain the neutron chain fission reaction.

There are two types of external neutron sources that have been suggested for this. The first is an accelerator-driven system in which accelerated deuterons impinge on a spallation neutron target embedded in the TRU fuel assemblies within the fission reactor. The second type is a fission–fusion hybrid reactor in which the 14.1-MeV neutrons from deuterium–tritium fusion reactions stream into TRU fuel assemblies blanketing the outside of the plasma chamber.

The subcriticality of a source-driven system provides two distinct advantages. As the fissionable TRU are depleted by burnup and their reactivity decreases, the fission chain reaction can be maintained at the same power by increasing the neutron source strength, enabling an increased fuel burnup to the radiation damage limit before the fuel is removed from the reactor. The second advantage is an increased reactivity safety margin to a prompt critical power excursion. For a critical reactor, this safety margin is $\Delta\rho \approx \beta$, the delayed neutron fraction, which is of order $\beta = 0.006$ for U235, but only order $\beta = 0.002$ for Pu239. On the other hand, for a subcritical reactor the safety margin is the subcriticality $\Delta\rho \approx \Delta k_{sub} = 1 - k \gg \beta$.

Argonne National Laboratory (ANL) designed IFR to be passively safe. SABR makes use of the same Na pool, metal-fueled technology as IFR, but there are several design differences between SABR and IFR. IFR is a single-core, critical reactor. (The most recent version of) SABR is a subcritical, source-driven reactor with ten physically separate but neutronically coupled cores. A change in the conditions in one core may affect the conditions of the neighboring cores in different ways. We describe here a dynamic safety analysis of this latest SABR#2.

A nodal neutron dynamics model is used to calculate the evolution of the power in the ten neutronically coupled cores which constitute SABR#2. The sub-criticality safety margin is $\Delta\rho \approx \Delta k_{\text{sub}} = 1 - k \gg \beta$, as compared to the usually much smaller subcriticality margin β = delayed neutron fraction of a critical nuclear reactor.

Modeling the coupled cores of the SABR. COMSOL Multiphysics is used as the platform of the dynamics model. The COMSOL model is a combination of an ITER-like fusion neutron source and an integral fast reactor (IFR)-like fast reactor. The annular fission reactor lies just outside the toroidal plasma chamber (as shown in many figures in this book). It is composed of ten separate sodium pools, each with its own fission core and heat exchanger. One of the pools is shown in Figure 27.2. Each pool contains a metal-fueled, sodium-cooled, pool-type, IFR-like fast reactor. There are 25 different physics packages available in the computation, several of which include solvers for fluid flow-through pipes, fluid and solid heat transfer, and structural mechanics. COMSOL can couple to other programs like MATLAB, which is used to couple the neutron kinetics calculations to the thermal hydraulic calculations. Figure 27.3 shows the overall computational flow of the model.

The MATLAB nodal kinetics solver uses tables of neutron kinetics data pre-computed by MCNP.

Define

$n_j(t)$ = neutron density of node j
β_j = delayed neutron fraction

For each time step in a transient, COMSOL sends the pool and node temperatures to the MATLAB kinetics solver. The MATLAB kinetics solver uses those

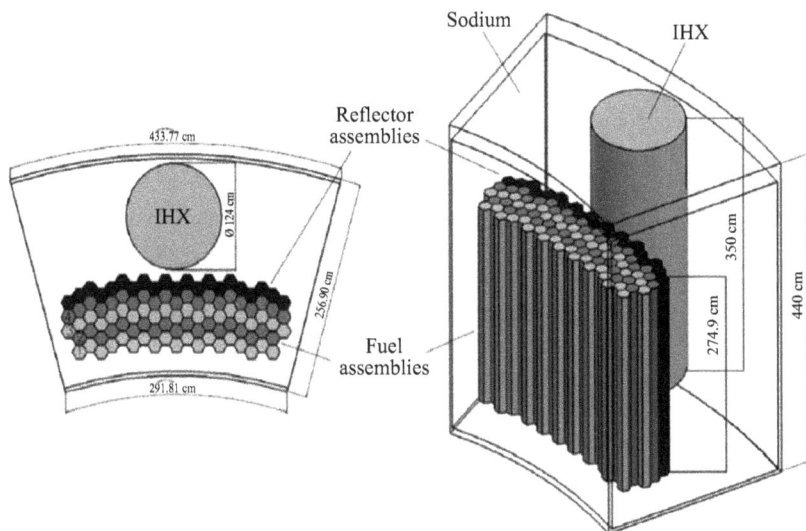

Figure 27.2   SABR#2 sodium pool

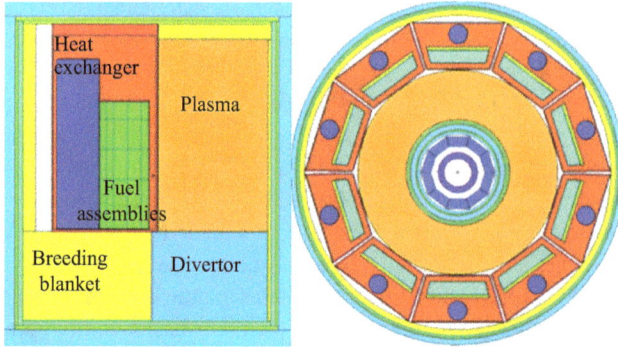

Figure 27.3   Side and top-down views of SABR#3 3-D MCNP model

temperatures to determine the change in fission power levels for each node and sends them back to COMSOL. COMSOL then uses the new power levels to calculate the new node $l_{e,j}$, $l_{a,j}$ = neutron leakage and absorption life-times, respectively, and pool temperatures for the next time step, etc.

27.1   Neutron kinetics model

The dynamics model uses a nodal neutron kinetics approach to calculate the time-dependent powers in each of the sodium pools produced by the fission neutrons plus $S_{\text{fus},j}$ = source neutrons contributed by the fusion plasma.

The $n,2n$ reactions are a non-negligible source in the system due to the hard neutron spectrum.

A node k to node j coupling coefficient α_{kj} represents the neutronic coupling of the ten nodes; α_{kj} is the probability that a neutron emitted from the surface of node k will impinge upon the surface of node j before entering any other node or being lost from the system. The rate of neutrons entering node j from node k is α_{kj} times the rate neutrons are escaping from node k, $(n_k(t)/l_{e,k})$.

This model does not contain explicit terms for the Doppler/sodium/fuel/clad reactivity coefficients. Instead, the terms Λ_{fj}, Λ_{2nj}, $l_{e,j}$, $l_{a,j}$, $S_{\text{fus},j}$, and α_{kj} collectively represent the Doppler/sodium/fuel/clad reactivity coefficients. How these terms are calculated is discussed in Sections 27.1.1 and 27.1.2. The nodal kinetics model constitutes a set of 70 coupled, stiff differential equations. We use the MATLAB program NodalSolve to numerically solve this system of equations for a given time step. NodalSolve uses MATLAB's intrinsic ode15s solver, which efficiently solves ordinary, stiff differential equations.

27.1.1   Calculation of nodal kinetics terms

MCNP6 calculates each of the nodal kinetics terms using a three-dimensional (3-D) model of the SABR#3. Figure 27.3 shows cross sections of the 3-D model.

The model contains ten sodium pools in addition to ITER-like components such as the vacuum vessel, plasma chamber, divertor, breeding blanket, and central solenoid. The ITER-like components outside the pools are homogenized for simplicity. Inside the pools, the individual subcomponents are homogenized separately. The heat exchangers are represented as homogenized cylinders. The pool vessels are modeled as hollow trapezoidal prisms. Any space that is inside the pool vessels but outside the cores and heat exchangers is by definition sodium. Each of the cores is divided into 125 cells and then homogenized. It is important to note that the cores are subdivided before being homogenized, not after. After the cores are subdivided and homogenized, they are referred to as nodes.

Different source and tally definitions are used depending on which kinetics terms are being calculated. To calculate Λ_{fj}, Λ_{2nj}, $l_{e,j}$, and $l_{a,j}$, an isotropic, uniformly distributed, volume neutron source is created in the cell labeled "Plasma" in Figure 27.4. The energies of the source neutrons are assigned using the energy spectrum for deuterium–tritium fusion neutrons included in MCNP6.

In each of the 125 cells in each of the ten nodes, tallies are placed for flux, absorption, fission, and $n,2n$ reactions. Direction-binned surface tallies are also placed on the outer surface of each node. The tallies considered to be of high relevance pass MCNP's ten statistical checks. For example, the absorption tallies in the sodium and the fission tallies in the fuel pass all of the tests. The resulting output file is quite large and contains results for over 5,000 separate tallies. A macro-enabled spreadsheet is used to automatically import the raw tally data from the MCNP output file. The spreadsheet condenses the data and calculates the effective kinetics parameters for each of the ten nodes.

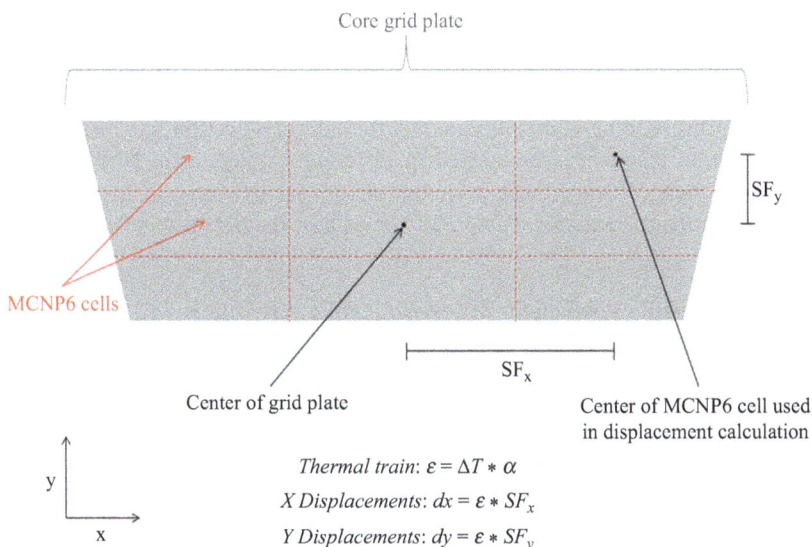

Figure 27.4 Diagram of displacement calculation for grid plate expansion

To calculate $S_{\text{fus},j}$, the same source definition as before is used (the isotropic, uniformly distributed, volume neutron source in the plasma cell), but the tallies are changed. A direction-binned surface tally is applied to the surface of each node, and the importance of each node is set to 0. Any plasma neutrons entering a node are tallied and immediately destroyed. This prevents the surface tally from double or triple counting any neutrons. Sixty million particles are run, and the tallies pass all statistical checks. This gives good enough statistical accuracy. There are only ten tallies in the output file, so an automated spreadsheet is not required.

A bit more effort is required to calculate the coupling coefficients a_{kj}. It is a two-part process. Part one uses the first source described: the isotropic fusion neutron source. Instead of tallies, the surface source writer in MCNP6 is used. The surface source writer stores the position, direction, weight, and energy of every particle crossing a designated surface. The surface source writer is applied to the surface of node 1. Doing so enables the recording of the steady-state inward and outward neutron fluxes of node 1. Ten million particles are run. For part two, the recording is used as a surface source on the surface of node 1. The importances of all nodes are set to 0, and surface tallies are set on the surface of each node. This ensures the tallies do not count a neutron more than once. Symmetry ($a_{kj} = a_{jk}$) is assumed, so it is only necessary to calculate this for node 1. One hundred million particles are run. There are only ten tallies in the output file, so no automation is required.

27.1.2   Calculation of feedback effects

The MCNP model calculates how various perturbations affect the terms of the nodal kinetics equations. The perturbations considered are core grid plate expansion, fuel axial expansion, fuel bowing, thermal Doppler broadening, and sodium expansion and voiding. The core grid plate expansion, axial expansion, and fuel bowing perturbations are manifested as changes in geometry. The sodium expansion and voiding is manifested as sodium density changes in the nodes and pools. The Doppler broadening is manifested as different cross-section sets that have had their absorption resonances broadened at different temperatures. Node and pool temperature changes cause these perturbations to occur, which in turn affect the terms in the kinetics equations. Table 27.1 shows how the parameters Λ_{fj}, Λ_{2nj}, $S_{\text{fus},j}$,

Table 27.1   Kinetics coefficients at several node temperatures

Kinetics coefficient	Node j temperature			
	529 K	829 K	1,129 K	1,429 K
Λ_{fj} (s)	4.289E−07	4.299E−07	4.284E−07	4.272E−07
Λ_{2nj} (s)	3.672E−04	3.480E−04	3.478E−04	3.472E−04
$S_{\text{fus},j}$ (cm^{-3}·s^{-1})	1.973E+11	2.009E+11	2.046E+11	2.083E+11
$l_{e,j}$ (s)	2.638E−07	2.598E−07	2.592E−07	2.586E−07
$l_{a,j}$ (s)	5.729E−07	5.757E−07	5.734E−07	5.717E−07

$l_{e,j}$, and $l_{a,j}$ change with node temperature. It is assumed the parameters in Table 27.1 are functions only of node j's temperature and are not affected by the temperature in adjacent nodes. Several calculations showed that while node j is affected by temperature changes in adjacent nodes, the effect is at least one order of magnitude smaller than the effect of a temperature change in node j. Table 27.2 shows how the coupling coefficients change with pool temperature. Note that the fusion neutron source into each node increases with the temperature of the sodium in each node because the intervening sodium density decreases.

For any change in temperature, the MCNP model simultaneously incorporates the effects of all the perturbation types mentioned earlier. This is why each node is subdivided into 125 cells and their materials homogenized. The subdivision and homogenization allow for changes in geometry and material properties to be modeled easily. The material density and composition of each of the 1,250 cells in the model are modified to reflect whatever change is desired. The MATLAB program UPM automates this process. UPM takes as input the temperature of each node, the temperature of each pool, and the physical displacements of each cell (i.e., geometry changes due to fuel bowing, grid plate expansion, or axial expansion). It calculates the new material densities and compositions for every cell in the model and writes this information to a text file in MCNP input format. It is easy to generate new MCNP input files for even the most complex combinations of perturbations. The following paragraphs describe how this MATLAB program accounts for each perturbation type.

UPM uses the core temperature to choose the cross-section set to be used for that node's materials. Doppler-broadened cross-section sets for temperatures of 600, 829, 1,200, and 1,500 K have been created. They were produced using the NJOY and the ENDF 7.0 libraries. These temperatures are slightly different from the temperatures used in Tables 27.1 and 27.2, but the difference is small.

In SABR, sodium voiding refers to two different things: voiding in the core and voiding in the pool. The sodium in the core is likely to change temperature and

Table 27.2   Nodal coupling coefficients at several pool temperatures

Coupling coefficient	Pool 6 temperature			
	529 K	829 K	1,129 K	1,429 K
$\alpha_{6,1}$	0	0	0	0
$\alpha_{6,2}$	0	0	0	0
$\alpha_{6,3}$	0	0	0	0
$\alpha_{6,4}$	0.00057	0.00059	0.00061	0.00064
$\alpha_{6,5}$	0.02753	0.02804	0.02865	0.02936
$\alpha_{6,6}$	0.74210	0.73755	0.73236	0.72640
$\alpha_{6,7}$	0.02753	0.02804	0.02865	0.02936
$\alpha_{6,8}$	0.00057	0.00059	0.00061	0.00064
$\alpha_{6,9}$	0	0	0	0
$\alpha_{6,10}$	0	0	0	0

density at a different rate than the sodium in the pool. Sodium voiding in the core affects the absorption and leakage terms of the nodal kinetics model. Sodium voiding in the pool affects the coupling and external source terms. UPM incorporates these changes by scaling the core and pool sodium densities separately based on their separate temperatures.

Before MCNP can determine how changes in geometry affect the kinetics terms, it needs to be understood how changes in temperature affect the geometry.

Simple hand calculations are used to determine the physical displacements caused by grid plate expansion and axial expansion. It is assumed the core grid plate expands and contracts isotropically with changes in temperature. The change in temperature is multiplied by the coefficient of thermal expansion to find the strain. To find the x-dimension displacements of a cell, the strain is multiplied by a scale factor. The scale factor is the distance between the x-coordinate of the center of the grid plate and the x-coordinate of the center of the cell in question. This is done for every cell in the model. The process is repeated for the y-dimension.

For the case of axial expansion, the same method that was used with the grid plate expansion is employed. The strain is calculated and multiplied by a scale factor. Here, the scale factor is the distance from the bottom of the core to the center of the cell in question. This gives the z-dimension displacements.

UPM uses these displacements to simulate material relocation. It does this by looping over each cell in the model and adding/removing materials where appropriate. The actual grid geometry of the MCNP model never changes; only the material compositions and densities of the node cells change. UPM accounts for fuel bowing in the same way it accounts for grid plate expansion and axial expansion. It uses displacements to simulate material relocation. The difference here is that the displacements are much harder to calculate. Fuel bowing is a highly non-linear process and requires a sophisticated model.

Fuel bowing is quantified in SABR using the Structural Mechanics Module in COMSOL Multiphysics.

This is a separate COMSOL model that is not part of the dynamics model. This model calculates the fuel assembly displacements in a single core at various power-to-flow ratios. The model uses a half-core geometry as shown in Figure 27.5 with a symmetry boundary condition to reduce computation time. The geometry contains all of the fuel assemblies and SiC flow channel inserts for a single core. Fuel pins are not included because they do not appreciably affect fuel bowing. The triangular swept mesh shown in Figure 27.6 is used on both the ducts and the inserts. The model accounts for geometric non-linearity, duct-to-duct contact, and duct-to-insert contact. A constrained boundary condition is applied to the bottom of the ducts. This is a simplifying assumption used for now because SABR's grid plate has not been designed yet. When more detail is available, the bottom boundary condition will need to be updated to a more realistic stiffness because the stiffness of the bottom boundary condition can have a significant impact on the fuel assembly displacements.

A 3-D temperature distribution is assigned to the mesh using the MATLAB program comFBtemp. It calculates the temperature at every mesh point on the inner surfaces of the ducts and the SiC inserts by assuming it is the same temperature as the

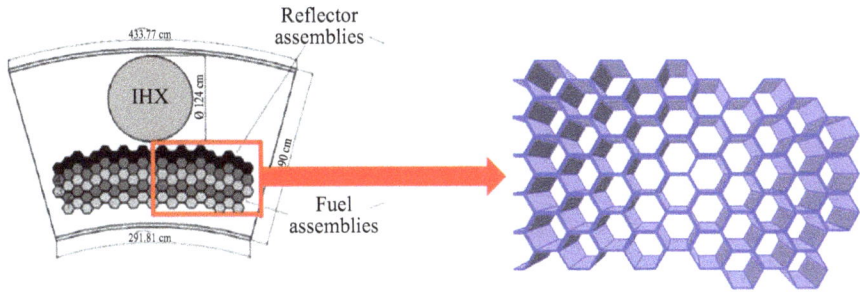

Figure 27.5 Geometry of fuel bowing model

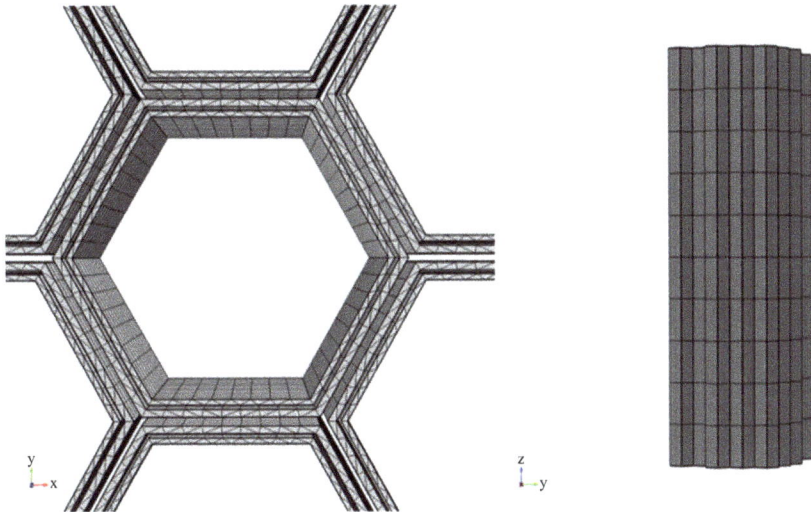

Figure 27.6 Top-down and side views of fuel bowing mesh

sodium adjacent to it. The program takes in the *x*-, *y*-, *z*-coordinates of each mesh point and calculates the temperature of sodium by integrating the amount of energy added to it as it travels up through the core in the *z*-direction. This requires the 3-D power distribution of the core. The MCNP model of SABR can produce exactly that by using FMESH tallies. This has nothing to do with the nodal kinetics model; it is just an additional task the MCNP model can perform. The core is broken into 5,301 cells, and the fission reactions in each cell are tallied. A 3-D power distribution is constructed using the tally results. A single power profile is calculated for the reference case. It is assumed the profile does not change during a transient.

The fuel bowing displacements are calculated for a model. It is a loop composed of sodium-filled pipes. Various boundary conditions simulate heat transfer from the core, across the heat exchanger, and into the sodium pool. A diagram of one of the ten loops in our model is shown in Figure 27.7.

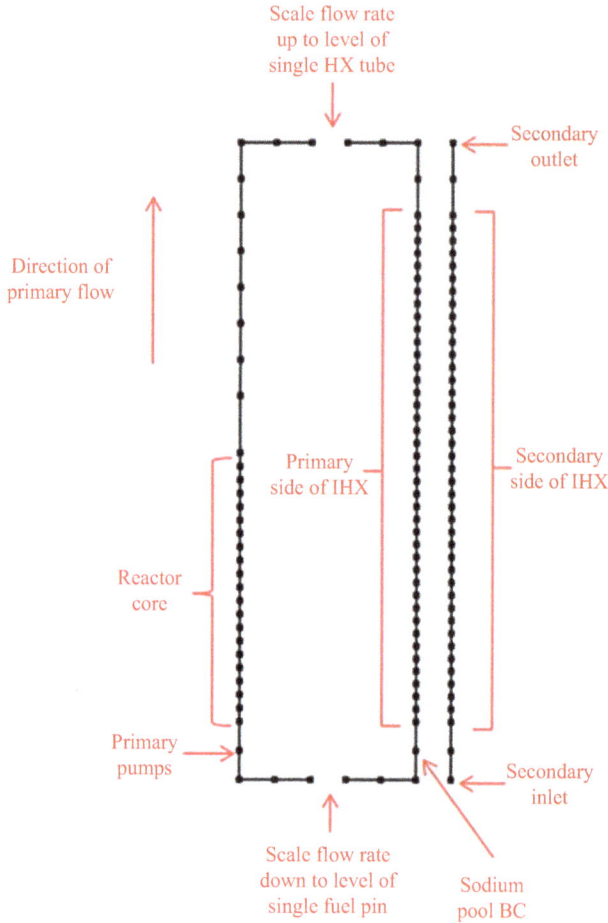

Figure 27.7   Diagram of sodium loop

27.1.3   Modeling the core

For each core, a single characteristic fuel pin is modeled. Two-dimensional axisymmetry is assumed as shown in Figure 27.8. The nodal kinetics model sets the power level in each of the ten different characteristic pins, and the Heat Transfer in Solids Module calculates the temperature distribution across the pins and cladding by solving the conduction equation for the case of 2-D axisymmetry. This is estimated as the highest ratio SABR can handle before design failure. The displacements for the z-dimension are not considered in the fuel bowing model because they are accounted for in the axial expansion perturbation model.

The fuel bowing for the most severe case possible is only a few millimeters. When these displacements are input to the MCNP model, no appreciable changes are seen in the kinetics parameters. Because of this, fuel bowing is not accounted for when performing the dynamic safety analysis of SABR.

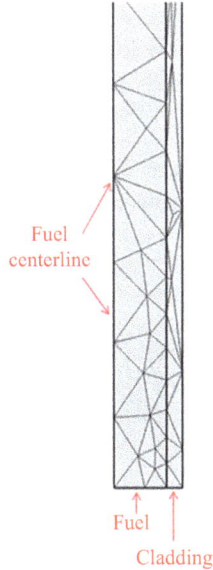

Figure 27.8   Geometry of fuel pin

27.1.4   Thermal-hydraulic model

The thermal-hydraulic component of the dynamics model utilizes COMSOL's Pipe Flow Module and Heat Transfer in Solids Module. Each of the ten pools is modeled separately. When modeling a pool, the entire primary loop (including the temperature distribution in the fuel, cladding, and primary heat exchanger) is considered, and only a small part of the secondary loop is considered. In the secondary loop, only the flow of sodium through the primary heat exchanger is modeled. Inlet and outlet boundary conditions are placed just outside the heat exchanger's secondary-side inlet and outlet. It is assumed the remaining power conversion and heat rejection cycles match those of a typical pool-type, sodium-cooled fast reactor design.

The Pipe Flow Module is used to create what is essentially a loop-type model of a SABR core. A special boundary condition in the loop accounts for additional heat capacity of the sodium pool. The structure of the loop-type model is similar to a RELAP5.

It also solves the continuity, momentum, and energy equations for the 1-D case:

where

ρ = density
C_p = specific heat capacity at constant stress
T = absolute temperature
u_{trans} = velocity vector of translational motion
q = heat flux by conduction

q_r = heat flux by radiation
α = coefficient of thermal expansion
S = second Piola–Kirchhoff stress tensor[11]
Q = heat source

For every point shown in Figure 27.8, the Pipe Flow Module calculates the heat transfer to the sodium coolant by solving the convection heat transfer equation:

where

q^{00} = heat flux
h = heat transfer coefficient
A = surface area
T_s = surface temperature
ρ = density
$\bar{u}$ = velocity vector
p = pressure
$\bar{\tau}$ = viscous stress tensor
F = volume force vector
c_p = specific heat capacity at constant pressure
T = absolute temperature
q = heat flux vector
Q = heat source
$\bar{S}$ = strain-rate tensor

This approach requires various empirical correlations that are referenced in Section 27.1.7.

27.1.5 Modeling the heat exchanger

The same procedures and equations described above are used to model the heat transfer in the heat exchanger. A single heat exchanger tube is modeled with the assumption of 2-D axisymmetry as shown in Figure 27.9. The Pipe Flow Module calculates the surface temperatures on both sides of the tube. The Heat Transfer in Solids Module uses the surface temperatures to calculate the temperature distribution and heat transfer across the tube. This gives the inlet and outlet temperatures for both sides of the heat exchanger.

$m_{pool,1}$ = mass of the pool at the beginning of the time step
$m_{pool,2}$ = mass of the pool in the middle of the time step
$E_{pool,1}$ = internal energy of the pool at the beginning of the time step
$E_{pool,2}$ = internal energy of the pool in the middle of the time step
$\dot{m}_{in}$ = mass flow rate of incoming sodium
ΔT = size of the time step
$c_{p,in}$ = specific heat capacity evaluated at T_{in}

In each pool, there are 5,700 tubes in the heat exchanger.

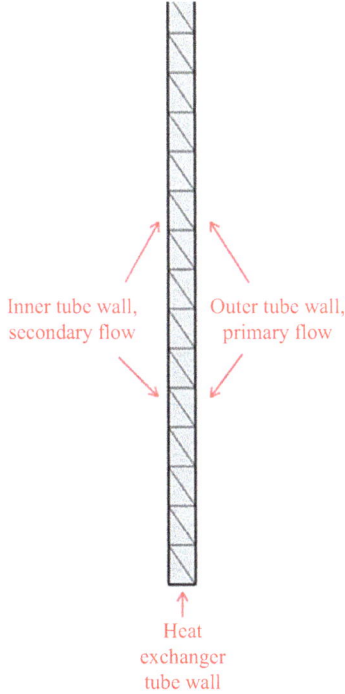

Figure 27.9   Heat exchanger tube geometry

The second step assumes thermodynamic equilibrium in the pool and calculates its average temperature with Eq. (11), which is the temperature of sodium exiting the pool and entering the core:

There are 37,520 fuel pins in the core. One of each is being modeled, so the mass flow rates are scaled proportionally to accurately model the system. For the core side of the loop, the flow rate is scaled to a value corresponding to single fuel pin. This is 1/37,520th of the core's primary mass flow rate. For the heat exchanger side of the loop, the flow rate is scaled to a value corresponding to a single heat exchanger tube. This is 1/5,700th of the core's primary flow rate. The location at which the flow rates are scaled is shown in Figure 27.10.

27.1.6   Modeling the sodium pool

The additional heat capacity of the sodium pool is accounted for by using the MATLAB program PoolTemp. As sodium in the primary loop exits the heat exchanger, the Pipe Flow Module calls PoolTemp to bring the sodium to thermodynamic equilibrium with the current pool temperature. This new temperature is set as the inlet temperature for the core. This is a three-step process.

Step one increases the mass and energy of the pool as sodium enters it from the primary outlet of the heat exchanger. Equations (9) and (10) are used to do this:

where c_p is the specific heat capacity evaluated at T_{in} and T_{out} is the temperature of outgoing sodium.

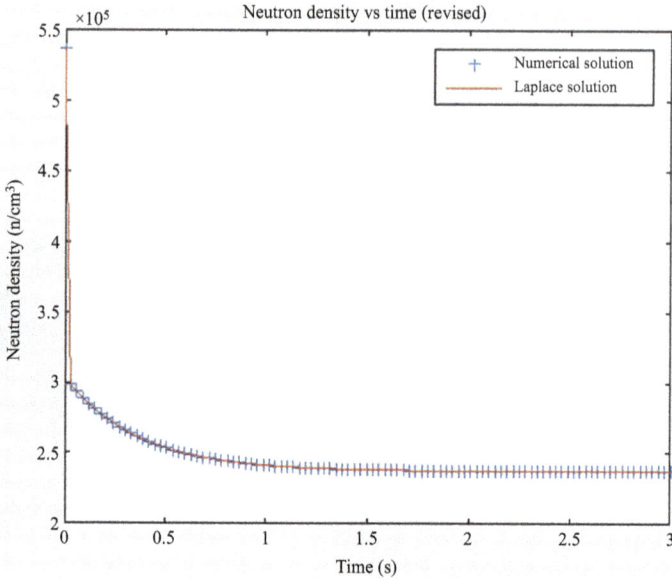

Figure 27.10   *Neutron density versus time (one-group delayed neutrons, 50%*
reduction in source strength at t = 0)

The third step decreases the mass and energy of the pool as sodium leaves it to enter the core inlet.

27.1.7   *Thermal property data and empirical correlations*

COMSOL is given the material properties for the fuel, cladding, and sodium as well as the empirical correlations for the friction factor and Nusselt number. It is assumed the properties of the fuel, cladding, and heat exchanger tube are constant. The same data used for the heat exchanger tube are used for the cladding.

The ANL sodium property correlations are used.

For the friction factor, the Zigrang–Sylvester approximation of the Colebrook–White correlation is used.

27.1.8   *Calculation of nodal heat transfer terms*

The kinetics model requires both node and pool temperatures as input. A node is defined as the fission core in each pool. This includes all fuel and reflector assemblies but not the heat exchanger and the surrounding sodium in the pool (Table 27.3).

COMSOL calculates T_{fuel}, T_{clad}, and T_{sodium} by finding the average temperatures in the fuel, clad, and sodium domains. It is assumed that $T_{reflector}$ is the same

Table 27.3   Fuel and clad property data

	TRU fuel	Oxide dispersion-strengthened cladding
Thermal conductivity (k)	10 W/m K	30 W/m K
Density (ρ)	3,861 kg/m^3	7,692 kg/m^3
Specific heat capacity (c_p)	738.15 J/kg K	650 J/kg K

temperature as the inlet temperature of the sodium because there is negligible heating in the reflector region. The masses of each component are hand-calculated constants that are hard coded into the COMSOL model.

The pool temperature is calculated using Eq. (11). Thermodynamic equilibrium is assumed in the pool indicating that $T_{pool} = T_{out}$.

27.2   Verification tests

SABR's dynamics model was subjected to several verification tests. No experimental data for SABR exist of course, so validation tests were unable to be performed. The verification tests were performed on individual subcomponents of the dynamics model: the neutron kinetics solver, thermal-hydraulic solver, and fuel bowing model. The dynamics model itself cannot be verified because there is no other model to compare it with. It is assumed that verification of the individual subcomponents of the dynamics model will suffice.

27.2.1   Neutron kinetics model

The MATLAB program NodalSolve is used to solve the nodal neutron kinetics equations over any desired time step. The nodal equations represent a stiff system because they are governed by multiple time constants that are different orders of magnitude: prompt neutrons, which change on the order of fractions of microseconds, and delayed neutrons, which change on the order of seconds.

The accuracy of the numerical solution was verified by taking a Laplace transform of the system and comparing it to the numerical solution. The system is too complicated to be easily solved with Laplace transforms, so it was simplified, and Mathematica was used to apply the transformations.[17] These tests were performed on single-node systems with a single delayed neutron group whose decay constant was varied from case to case. For brevity, only a single case is shown.

The source term is reduced by 50% at $t = 0$, and the resulting Laplace solution is calculated.

Figure 27.10 compares the two solutions, which agree quite well, providing verification for the neutron kinetics solver.

27.2.2   Thermal-hydraulic model

RELAP5 (see [18] for an introduction of RELAP5) was used to test the thermal-hydraulic component of our model. Because RELAP5 cannot handle the complexity of the SABR model, it was simplified to a single-node, loop-type core with no feedback. The remaining core geometry, composition, and thermal-hydraulic conditions are identical to the SABR design. The steady-state temperatures and pressure drops were compared, and the transient maximum fuel temperatures and maximum coolant temperatures were compared for every type of accident scenario in the dynamic safety analysis.

The steady-state temperature comparisons are shown in Table 27.4. The temperatures and pressure drops agree well with RELAP5.

For the sake of brevity, this paper shows only one of the six transient cases compared. The maximum fuel and coolant temperatures for a 50% loss-of-flow accident (LOFA) in core 1 are shown in Figure 27.10.

The transient results agree well, and the thermal-hydraulic model is considered verified.

27.2.3   Fuel bowing model

COMSOL was used to model the structural mechanics component of the fuel bowing analysis. The fuel bowing model was verified by using it to model an International Atomic Energy Agency (IAEA) structural mechanics verification problem.[19] The verification model used similar settings for the mesh and physics solvers as in SABR's fuel bowing models.

The problem solution is a hand calculation made by the IAEA's International Working Group on Fast Reactors. The problem contains a single fuel assembly made of stainless steel, and it uses Cartesian coordinates with the origin centered on the bottom face of the duct. The z-axis is parallel with the duct axis. The duct wall is 3-mm thick with an across-flats dimension of 132.9 mm. The material properties are $T_{ref} = 20°C$, $\alpha = 18.6 \times 10^{-6}/°C$, $\nu = 0.3$, and $E = 170$ GPa. The temperature field is described by the piecewise function below with $a_0 = 212.5°C$, $a_1 = 0.500°C/mm$, $a_2 = 0.125°C/mm$, and $a_3 = -3.333 \times 10^{-4}°C/mm$.

Table 27.4   Steady-state parameters

	RELAP5	**COMSOL**
Average fuel temperature (K)	725	720
Maximum fuel temperature (K)	801	802
Core inlet temperature (K)	619	619
Core outlet temperature (K)	759	759
Secondary inlet temperature (K)	590	590
Secondary outlet temperature (K)	739	739
Core pressure drop (kPa)	70	71
Primary pressure drop (kPa)	264	266
Secondary pressure drop (kPa)	33	33

Table 27.5 Comparison of displacement estimates

	COMSOL model	**IAEA problem**	**Percent difference**
x displacement	−11.66 mm	−12.36 mm	5.7%
z displacement	32.91 mm	31.86 mm	3.2%

A similar swept, triangular mesh is applied to the model. The same minimum/maximum parameters as the SABR fuel bowing model are used, but the axial mesh is refined a bit. The axial temperature profile in the problem varies more strongly than the axial temperature profile of SABR. The same physics solvers are used as in the SABR model, and geometric non-linearity is included. The *x* and *z* displacements at the top of the assembly are calculated. The results are shown in Table 27.5.

The results agree well, and the fuel bowing model of SABR is considered to be verified.

27.3 Dynamic safety analysis

A dynamic safety model was developed and applied to analyze the safety characteristics of the SABR.

The temperatures shown in Table 27.4 are used as the initial conditions for temperature. A maximum time step of 0.1 s was used for every accident scenario with a total time of 125 s unless more time was needed; 125 s is the minimum time required for the coolant flow rates to coast down and reach a constant value. SABR uses electromagnetic sodium pumps whose power source is connected to a mechanical flywheel system. When power is lost, the pumps have a 10-s halving time. The COMSOL dynamics model uses these data to control the primary and secondary flow rates as a function of time.

When both pumps fail, the flow rate does not coast down all the way to zero. There is some natural circulation in the core. The method prescribed by Todreas and Kazimi was used to estimate natural circulation in the primary loop[20]:

g = acceleration due to gravity
ΔL = height difference between the center of the core and the center of the heat exchanger
ρ_0 = reference density of sodium
c_p = average heat capacity of sodium
n = 0.2 for turbulent flow
R = hydraulic resistance coefficient

This natural convection model assumes steady-state conditions. While the safety analysis is not simulating steady-state conditions, this model can still be used. Any accident results in either failure or new steady-state conditions. If the accident results in new steady-state conditions, the use of the model is justified. If the accident results in failure, use of the model is irrelevant. This dynamic safety analysis is only interested in determining if and when failure occurs, not what happens after failure. The model predicts natural circulation through the rate for full power and decay power levels, respectively. The decay power for SABR's current design has not been explicitly calculated. It is conservatively assumed that the decay power is a constant 7% of full power. This assumption is based on studies of decay power in TRU fuel.[21] Each case in the dynamic safety analysis was evaluated on a pass/fail basis. A case failed if the coolant reached its boiling point of 1,156 K or the fuel reached its melting point of 1,473 K. Unless otherwise noted, all accident-initiating events were run in core 1 only. Cores 2 through 10 remained unaffected.

27.3.1   *Accident scenarios and corrective actions*

The accidents we simulated were loss-of-flow accidents (LOFAs), loss-of-heat sink accidents (LOHSAs), and loss-of-power accidents (LOPAs). The primary and secondary loops of each SABR pool have two pumps, so a 50% LOFA/LOHSA simulates the loss of one pump in the primary/secondary loop while 100% LOFA simulates the loss of both pumps in the primary loop. In core 1, 50% LOFAs/ LOHSAs and 100% LOFAs/LOHSAs/LOPAs were run with no corrective action taken. These same cases were also run with two different corrective actions: fusion plasma shutdown and control rod insertion.

SABR's plasma is maintained by external power sources that heat the plasma and drive confining current in the plasma. Simply switching off these external power sources would reduce the neutron source strength as the plasma cooled down. The energy confinement time would decrease as the plasma current and plasma heating decreased.

To date, SABR's design does not explicitly include control rods, so a rod bank similar to the one in ANL's Advanced Burner Test Reactor (ABTR) is used. The ABTR uses a bank of ten control rods that provide up to −40.19$ of reactivity. The ABTR is a good analog because it has a power, fuel composition, and delayed neutron fraction comparable to SABR. A SABR core generates 300 MW(thermal) of power while the ABTR generates 250 MW(thermal). SABR's fuel is TRU while the ABTR's fuel is a mixture of TRU and uranium. SABR's delayed neutron fraction is $\beta_{\mathrm{eff}} = 0.0030$ and ABTR's is $\beta_{\mathrm{eff}} = 0.0033$.

If the rod distribution shown in Figure 26.17 is used, nine control rods can be fit into each SABR core. Each of the ten control rods in the ABTR have different reactivity worths, but here the average is taken, and it is assumed each rod has the same worth. Under that assumption, each rod would be worth 4.02$. Equation (26) is used to convert that to a reactivity of $\Delta\rho = -0.0134$. If the confinement time in the plasma is on the order of seconds, the neutron source strength might be expected to decay exponentially with a time constant on the order of seconds. There are constraints on ramping down magnetic fields that might lead to the necessity of slower neutron source decays. The Fusion Research Center at Georgia Institute of

Technology (Georgia Tech) is currently examining this issue. Initial results suggest the plasma power can be reduced to 0 within 10 s. For the purpose of the present work, the conservative assumption of an exponential decay with a time constant of 10 s is made.

Most fast reactor control systems initiate shutdown when certain parameters such as core temperature or power-to-flow ratio reach a certain value. Once the threshold is reached, the control system takes on average 2 s to respond. For SABR, our threshold is pump failure. When a pump fails, it is assumed the reactor control system takes 2 s to notice and another 3 s to respond by initiating a plasma shutdown. Thus, the plasma shutdown is initiated 5 s after a pump in either the primary or secondary systems in any of the fission cores stops.

When using control rods as a corrective mechanism, the plasma power remains at 100%, and the control rods are inserted to decrease the fission power several seconds after an accident begins. It is assumed it takes 2 s for the reactor control system to notice the pump failure and 1 s to initiate a control rod drop in all ten fission cores. It is assumed the control rods take 2 s to fully insert once they are released.

SABR's total control rod worth is estimated by dividing $\Delta\rho$ by SABR's β_{eff} and multiplying by nine rods. This yields a control rod worth of $-39.67\$$ per pool lowering k_{eff} from 0.92185 to 0.83047, which corresponds to a 15.985% decrease in the absorption lifetime l_a.

27.3.2 Accident analyses

SABR is passively safe (i.e., avoids failure without any external corrective action taken) against 50% LOFAs. No fuel melting or coolant boiling occurs, and the temperatures remain well below the threshold of failure. A similar pair of curves (not shown) demonstrates that SABR is passively safe against 50% LOHSAs as well.

Feedback effects would have a negligible effect on reducing the core power level in the 100% LOFA, control rod insertion would reduce the power level to 36% of full power, and shutting off the plasma neutron source would reduce the core power level to the decay heat level of 7% of full power. Similar power profiles were obtained for the 100% LOHSA and LOPA, with the exception being that the reactivity feedbacks reduced the core power level slightly more in the LOHSA case than with the 100% LOFA and 100% LOPA. SABR reaches the point of failure during a 100% LOFA with both coolant boiling and fuel melting regardless of any corrective action that is applied. When no corrective action is applied, reactivity feedbacks only slightly reduce the power level, and the core reaches the point of failure after 27 s. When control rods are inserted, they reduce the core power to 36% of full power. The reactor continues to operate at this reduced power level until it reaches the point of failure after 59 s. When the plasma is shut down, the plasma power drops to zero, and the reactor power level drops to the decay heat level of 7% of full power after 33 s, but the decay power is not adequately removed by natural circulation. The remaining decay power causes the coolant to boil and the fuel to melt at 208 and 275 s, respectively, after accident initiation.

SABR fails during the 100% LOHSAs even with corrective actions. The times to failure are longer than those for the 100% LOFAs because of the thermal capacitance of the pool. The more sodium there is in the pool, the longer is the time to failure. The reactivity feedbacks are slightly more impactful than for the 100% LOFAs, but they are still quite weak. The control rods and plasma shutdown mechanisms perform similarly as they did during the 100% LOFAs. The control rods do not provide enough negative reactivity to shut down the core. The plasma shutdown successfully shuts down the fission core and lowers the power level to decay heat levels before failure; however, the residual decay heat is not removed, and the reactor reaches the point of failure in about 30 min.

SABR fails for the 100% LOPAs with or without corrective actions. Here, a worst case is taken for the LOPA, namely, that the fission reactor and fusion neutron source have separate power supplies and only the fission reactor loses power. The results are almost identical to those of the 100% LOFAs because the reduction in primary flow effectively decouples the core from the heat exchanger and secondary cycle.

27.3.3   Neutronic coupling

As expected, the neutronic coupling among the cores is strong. The top plot in Figure 26.22 illustrates the effect of control rod insertion in core 5 on the power levels in each of the ten cores. Also shown for comparison are the core powers when control rods are inserted in all ten cores.

27.4   Discussion

A multi-pool dynamic safety model has been created for SABR#2 using COMSOL Multiphysics for the thermal hydraulics, a nodal neutron kinetics model with parameters evaluated using Monte Carlo, and feedback models accounting for the effect of temperature changes on cross sections and on fuel movement. The model was verified against RELAP5, analytical solutions, and benchmark problems. SABR has been found to be passively safe (no coolant boiling or fuel melting even without control action) in accident scenarios that model the failure of 50% of the coolant pumps in either the primary or secondary loops. For the failure of 100% of the pumps in either the primary or secondary loops, the power could be quickly reduced to the 7% decay power level by shutting off power to the plasma neutron source, but the low level of natural circulation was insufficient to remove the accumulating decay heat, and coolant boiling followed by fuel melting eventually took place. Future design work will address the decay heat removal in SABR.

Control rods were found to be a less effective shutdown mechanism than switching off the power to the plasma source, and reactivity feedbacks were found to be a relatively ineffective shutdown mechanism in a subcritical system.

Several years prior to this work, a dynamic safety analysis was run on an older version of SABR's design.[23] RELAP5 was used to run LOFAs, LOHSAs, and LOPAs. A point kinetics model was used, and the reactivity feedbacks accounted for were Doppler broadening and sodium voiding. The results were similar to those found in this paper, namely, that SABR could survive 25% pump failure accidents without control action, could survive 50% pump failure accidents in the secondary system (but not in the primary system) without control action, and could survive a LOPA because the neutron source also was turned off. The times to failure were marginally shorter than in the present work.

The results of this previous work were used to update SABR's design in 2014. There were two objectives of the redesign: (1) improve SABR's safety performance by changing it from a loop-type core to a pool-type core and (2) increase its practicality by more realistically integrating the fission and fusion reactor components together. Accomplishing the second objective meant changing the once annular and continuous fission core into ten separate cores. Doing so totally changed the neutron kinetics of SABR. Originally, it could be assumed that any perturbation would affect the core uniformly, and point kinetics were used. With SABR's fission core now composed of ten physically separate but neutronically coupled cores, that assumption no longer applies. Conditions among the cores could vary and give rise to standing power oscillations or power tilts. Thus, the nodal neutron kinetics model presented in this paper was developed to replace the point kinetics model.

In summary, *this paper differs from the work of the previous chapter in three major ways: (1) nodal kinetics was used to model a ten-core reactor instead of point kinetics being used to model a single-core reactor; (2) a pool-type reactor is modeled instead of a loop-type reactor; (3) more reactivity feedbacks are included in the kinetics model. The present work accounts for Doppler broadening, sodium voiding, fuel and reflector axial expansion, core grid plate expansion, and fuel bowing while the previous work only accounted for Doppler broadening and sodium expansion.* It is also useful to compare SABR's performance with that of the Experimental Breeder Reactor-II (EBR-II). The choice of metal fuel and Na-pool design for SABR was influenced by the passive safety demonstration tests performed in EBR-II in 1986. In those tests, EBR-II exhibited full passive safety during uncontrolled 100% LOFAs and LOHSAs. These EBR-II results are different from those calculated for SABR for two reasons: (1) SABR's neutron source and (2) SABR's lack of adequate decay heat removal.

SABR's operation with an external neutron source weakens the effect of reactivity feedbacks. In EBR-II, feedbacks generated negative reactivity that significantly affected the neutron balance. In SABR, the fission chain reaction is maintained by an external neutron source that offsets $\Delta k_{\mathrm{sub}} = 1 - k \geq 0.03$ of reactivity, so the much smaller reactivity effects of feedback are relatively less important than they are in EBR-II.

Furthermore, once reactivity feedbacks shut down the reactor it was possible to remove all decay heat through passive means. EBR-II incorporated additional heat exchangers whose sole purpose was to remove decay heat using natural circulation. SABR's current design has no explicit decay heat removal system, and that is the reason for fuel melting and coolant boiling even after SABR is successfully shut down by turning off the plasma. Additional decay heat removal will be included in the next iteration of the SABR design.

Chapter 28

The panacea of just harvesting "free" green energy?

Why not just harvest free green energy? Isn't it much simpler?

An appealing, but in large part misleading, argument has been made that "green" energy freely existing in nature—wind, solar, rivers, waves—can just be harvested to replace the power produced by burning carbon-based fuel. Many people argue wouldn't it be simpler, and somehow purer, to just use this "free" energy than to develop all of this nuclear energy?

There clearly is an enormous amount of natural energy on the planet, as witnessed by Niagara Falls, the warming of England and Europe by the flow of the Gulf Stream, the enormous energy in hurricanes and monsoons, the temperatures in the Arizona desert, California forest fires, the waves and wind at sea, etc., and we can indeed tap into it. However, as usual, the devil is in the details. The reality with wind and solar energy, for example, is that the wind does not always blow and the sun does not always shine, in particular when and where energy is needed. Waves, of course, occur mostly in the oceans, where the electricity demand is modest.

The reality is that harvesting "free energy" would necessitate building and maintaining not only massive, many hundreds to thousands of square miles of solar collectors and windmills and wave energy collection systems to harvest this "free" energy, but would also require building many massive energy storage units to store wave, wind and solar-generated electrical energy from the times it is harvested until it is needed, and building many massive electricity transmission systems to get the power from where it is produced to where it is stored to where it is needed. Vast arrays of batteries or, geography permitting, vast artificial hydroelectric systems that use solar, wave and wind energy to pump water to higher elevations, from which it could be later released to produce hydroelectric energy when energy was needed, would have to be constructed for energy storage between the time of "harvesting" and the time of need. Such transfers of energy from one form to another are not usually very efficient.

For both solar and wind energy, there is the additional problem that the energy intensities of the solar radiation and the wind are quite small (implying that the space that must be devoted to collecting a large amount it must be quite large), and the efficiency of converting solar or wind power to electrical power is relatively low. There is also the visual pollution, bird deaths, and maintenance requirements etc. that would be associated with all those solar collectors and windmills (some taller than the Statue of Liberty), and power lines coating the earth, and the sea in the case of wind and wave energy. It is not a reassuring picture.

While rivers do tend to always flow, if not always at the same rate, the most promising hydro-electric locations (in the USA) were dammed in the last century. Similarly, geothermal hot springs tend to be relatively reliable, but the most likely sources in the USA have already been utilized for minimal power production.

There is also the detail of "load following". The demand for electricity varies throughout the day and night (as does the source of electricity in systems that must accommodate variable wind and solar sources of power). A coal or gas/oil-fueled or a nuclear power plant can simply adjust its operating power level to follow the demand, but a solar or wind power plant is unable to do this and, in fact, would exacerbate the "load following problem".

We believe that when the detailed analyses of harvesting, storing and transmitting wind, solar, wave, etc. produced energy are made and the psychological impact of destroying the beauty of the landscape is evaluated (and this should be done soon, by non-promoters), these costs are likely to be much greater than the costs of implementing clean abundant nuclear energy to displace carbon-burning fuels to provide the world's energy needs. We anticipate that such studies will show that nuclear is the most economical, reliable, environmentally benign, psychologically tolerable and practical choice for meeting the world's future clean energy needs.

Chapter 29
Summary, discussion, and recommendations

About 4.17 trillion kWh(e) of electricity were generated at utility-scale electricity generation facilities in the United States in 2018. Of this, 64% was from fossil fuels (coal, natural gas, petroleum, and gas), 19% was from nuclear energy, and about 17% was from solar, wind, hydro, and other "renewable" energy sources. The task before us is to displace this 64% of the electricity from fossil fuels with electricity that does not put more carbon in the atmosphere and that minimizes negative environmental impact, and to make similar displacements elsewhere in the developed and developing world.

We have documented that environmentally benign nuclear power (fission today, supplemented by fusion later in the second half of this century) is a technically viable replacement for carbon-based fuel (coal, oil, gas) burning power, today. We have calculated that the nuclear fission fuel resources are sufficient to meet the world's total needs for electric power for several decades, even operating on the present inefficient "once-through" nuclear fuel cycle. For the longer term, subcritical fusion–fission breeder reactors driven by fusion neutron sources, and later pure fusion reactors, can provide mankind's total electrical energy needs into the indefinite future.

Nuclear fusion technology is being developed for power production in the second half of the present century and beyond, and nuclear fusion fuel resources (lithium and seawater) are adequate to provide mankind's power production needs for the indefinite future.

We believe that nuclear (fission plus fusion) power is the most realistic, reliable, and environmentally acceptable replacement for burning carbon-based fuel to produce electric power on the large scale required, in the places and at the times power is needed. We therefore recommend that the industrialized and industrializing nations of the world join forces for the further development and worldwide implementation of clean nuclear power (fission now, fission supplemented by fusion in the near future, and fusion in the future) to displace power produced by the burning of carbon-based fuel.

We have demonstrated that the technically advanced breeding nuclear fuel cycle enabled by Fusion–Fission Hybrid reactors would enable the world's known uranium and thorium nuclear energy resources to provide the world's current electric power requirements for millennia; and that the technically advanced burning nuclear fuel cycle enabled by Fusion–Fission Hybrid reactors would resolve the "nuclear waste" problem.

Further reading: Georgia Tech Fusion–Fission Hybrid Papers

1. "Transmutation Facility for Weapons-Grade Plutonium Disposition Based on a Tokamak Fusion Neutron Source", Fus. Techn., 27, 326, 1995.
2. "A Tokamak Tritium Production Reactor", Fus. Techn., 32, 563, 1997.
3. "A Tokamak Tritium Production Reactor Design II", Fus. Techn. 33, 443, 1998.
4. "Capabilities of a DT Tokamak Fusion Neutron Source for Driving a Spent Nuclear Fuel Transmutation Reactor", Nucl. Fus., 41, 135, 2001.
5. "A Fusion Transmutation of Waste Reactor", Fus. Eng. Des., 63, 81, 2002.
6. "Nuclear and Fuel Cycle Analysis for a Fusion Transmutation of Waste Reactor", Fus. Eng. Des., 63, 87, 2002.
7. "A Fusion Transmutation of Waste Reactor", Fus. Eng. Des., 63, 81, 2002.
8. "A Fusion Transmutation of Waste Reactor", Fus. Sci. Techn., 63, 41, 2002.
9. "Comparative Fuel Cycle Analysis of Critical and Subcritical Fast Reactor Transmutation Systems", Nucl. Techn., 144, 83, 2003.
10. "Nuclear Design and Analysis of the Fusion Transmutation of Waste Reactor", Fus. Sci. Techn., 45, 51, 2004.
11. "A Superconducting Tokamak Fusion Transmutation of Waste Reactor", Fus. Sci. Techn., 45, 55, 2004.
12. "Subcritical Transmutation Reactors with Tokamak Fusion Neutron Sources", Fus. Sci. Techn., 47, 1210, 2005.
13. "A Subcritical, Gas-Cooled Fast Transmutation Reactor with a Fusion Neutron Source", Nucl. Techn., 150, 162, 2005.
14. "A Subcritical, Helium-Cooled Fast Reactor for the Transmutation of Spent Nuclear Fuel", Nucl. Techn., 156, 2006.
15. "Transmutation Missions for Fusion Neutron Sources", Fus. Eng. Des., 82, 11, 2007.
16. "Advances in the Subcritical, Gas-Cooled, Fast Transmutation Reactor Concept", Nucl. Techn., 159, 2007.
17. "Tokamak D-T Fusion Neutron Source Requirements for Closing the Nuclear Fuel Cycle", Nucl. Fus., 217, 2007.
18. "Fuel Cycle Analysis of a Subcritical Fast Helium-Cooled Transmutation Reactor with a Fusion Neutron Source", Nucl. Techn., 158, 2007.
19. "Advances in the Subcritical, Gas-Cooled, Fast Transmutation Reactor Concept", Nucl. Techn., 159, 72, 2007.

20. "Sub-Critical Transmutation Reactors with Tokamak Fusion Neutron Sources Based on ITR Physics and Technology", Fus. Sci. Techn., 52, 719, 2007.
21. "Tokamak Fusion Neutron Source for a Fast Transmutation Reactor", Fus. Sci. Techn., 52, 727, 2007.
22. "A TRU-Zr Metal-Fuel Sodium-Cooled Fast Subcritical Advanced Burner Reactor", Nucl. Techn., 162, 53, 2008.
23. "Georgia Tech Studies of Sub-Critical Advanced Burner Reactors with a D-T Fusion Tokamak Neutron Source for the Transmutation of Spent Fuel".
24. "Dynamic Safety Analysis of the SABR Subcritical Transmutation Reactor Concept", Nucl. Techn., 171, 123, 2010.
25. "Tutorial: Principles and Rationale of the Fusion–Fission Hybrid Burner Reactor".
26. "Principles and Rationale of the Fusion–Fission Hybrid Burner Reactor", AIP Conf. Proc., 1442, 31, 2012.
27. "Advanced Fuel Cycle Scenario Study in the European Context Using Different Burner Reactors", ISBN 978-92-64-9917, OECD, 2012, 351.
28. "Transmutation Fuel Cycle Analyses of the SABR Fission–Fusion Hybrid Burner Reactor for Transuranic and Minor Actinide Fuels", Nucl. Techn., 182, 2013.
29. "Resolution of Fission and Fusion Technology Integration Issues: An Updated Design Concept", Nucl. Techn., 187, 2014.
30. "The SABrR Concept for a Fission–Fusion Hybrid Fissile Production Reactor", Nucl. Techn., 187, 1, 2014.
31. "Solving the Spent Nuclear Fuel Problem by Fissioning Transuranics in Subcritical Advance Burner Reactors", Nucl. Techn., 200, 15,2017.
32. "Dynamic Safety Analysis of a Subcritical Advanced Burner Reactor", Nucl. Techn., 200, 250, 2017.
33. "Georgia Tech Studies of Sub-Critical Advanced Burner Reactors with a D-T Fusion Tokamak Neutron Source for the Transmutation of Spent Nuclear Fuel", J. Fusion Energy, 2019.

Glossary for "Fusion–Fission Hybrid Reactors"

Breeding	The neutron transmutation of a non-fissionable (or very weakly fissionable) atomic nucleus into a strongly neutron fissionable atomic nucleus.
Breeding Fuel Cycles	Fuel cycles that emphasize the neutron transmutation of non-fissionable U238 into fissionable Pu239 and Pu241 or the neutron transmutation of non-fissionable Th232 into fissionable U235 and U233.
Burning	The neutron (transmutation) fission of fissionable nuclei remaining in used nuclear fuel that has been removed from nuclear reactors.
Burning Fuel Cycles	Fuel cycles that emphasize fissioning the remaining fissionable material in used fuels that have been removed from nuclear reactors in order to maintain efficient reactor performance.
BWR	Boiling water (nuclear fission) reactor
CANDU	Canadian D2O pressure-tube (nuclear fission) reactor
Cross Section	Measure of the probability for a given type of neutron–nucleus reaction.
Decay Half-Life	The time in which one half a given mass of radioactive material will decay away.
Divertor	A magnetic configuration to achieve the diversion of plasma ions escaping the confinement region away to a remote "diverter region" to interact with the wall in order to inhibit wall-sputtered impurities from being introduced into the main plasma.
Deuterium	An isotope of hydrogen with one proton and one neutron in the nucleus.
Fast Reactors	Nuclear reactors in which most of the nuclear reactions take place in the 10–100 thousand eV energy range.
Fertile Isotope	A non-fissionable isotope that will turn into a fissionable isotope upon neutron capture.
Fission Chain Reaction	A fission event produces two to three neutrons, of which some will be lost from the reactor, some will be captured in non-fissionable materials and some will produce another fission. If, on average, the number of fission-produced

neutrons that cause a secondary fission is 1 then the reactor is *critical*, <1 the reactor is *subcritical*, or >1 the reactor is *supercritical*.

Fission Cross Sections   Probabilities for neutron fission of different fissionable materials.

Fission Products   The intermediate atomic mass nuclei which result from the fission of uranium and plutonium.

Fusion Cross Sections   Probabilities for different light ion pairs to undergo fusion.

Fusion–Fission Hybrid (FFH)   A nuclear fission reactor supplemented with a coupled nuclear fusion neutron source.

Greenhouse Gas   Part of the incident solar radiation is reflected from the earth's surface as thermal radiation with frequency nu = ν_{th} (nu-thermal) and energy $E = h\nu_{th}$ (h is Planck's constant). If gases are present in the atmosphere which have an atomic electronic structure such that the difference in energy levels between an occupied atomic electronic state and an unoccupied higher atomic electronic state ΔE closely corresponds to the frequency of thermal radiation, this thermal radiation will be reabsorbed and trapped in the earth's atmosphere. That part of the reflected thermal radiation that would otherwise have been lost into space from the earth's atmosphere is thereby trapped in the earth's atmosphere to heat it. Greenhouses work like this because the carbon dioxide given off by the plants has just the right atomic electronic structure, and there are other gases as well.

Half Life   The time in which half of the radioactivity present at a given time will decay away.

High Level Waste   Highly radioactive nuclear waste.

High-Level Waste Repository   A repository for secured thousand-year storage of high-level radioactive nuclear waste.

HTGR   High-temperature gas-cooled reactor.

IFR Integral Fast Reactor   A fast neutron spectrum nuclear fission reactor concept, which is the basis for Versatile Test Reactor being built is the United States.

Isotope   A nuclear species, e.g., U238.

ITER   International Tokamak (Experimental) Reactor which began operation in the 2020s.

Lawson Criterion   A measure of plasma energy and pressure confinement relative to that required for a fusion reactor.

LMFBR   Liquid metal fast breeding reactor, a fast spectrum nuclear fission reactor focused on the breeding fuel cycle.

Mirror Fusion Confinement	A magnetic fusion confinement concept.
Neutron Cross Section	The probability for a neutron to produce a given reaction with a given atom type (k = scatter, capture, fission, n-$2n$, etc.) can be written as the product of the neutron density n, the neutron speed V, the atom number density N, and the cross section for process "x", σ_x, i.e., as $nVN\sigma_x$.
Neutron Spectrum	The distribution in energy of the neutron population in a nuclear fission or fusion reactor.
Neutron Transmutation	Transmutation basically means "change." Neutron transmutation is change to the atomic nucleus caused by neutron absorption. When a neutron is absorbed into an atomic nucleus containing Z protons and A–Z neutrons it forms an atomic nucleus still with Z protons but now with $(A+1)$–Z neutrons.
Nuclear Fuel Cycle	The series of industrial processes that fissionable fuel goes through from mining to disposal.
Nuclear Power Reactor Accidents	Three-Mile Island, Chernobyl, Fukushima
Once-through fuel cycle	A nuclear fuel cycle in which the fuel remains in the reactor until it reaches the radiation damage or effectiveness limit, then is removed and stored or buried.
PRISM	A Na-cooled, pool-type critical fast reactor.
psi	Pounds per square inch.
PWR	Pressurized water (nuclear fission) reactor.
Radioactive Decay	Some atomic nuclei that are formed with atomic mass A (N = # neutrons + Z = # protons) are unstable and will decay by emission of a proton ($A = 1$, $Z = 1$), a neutron ($A = 1$, $Z = 0$), an alpha particle (2 protons + 2 neutrons), an electron, and or a photon. Such unstable nuclei are referred to as radioactive. The length of time over which one-half of the radioactive nuclei present will have decayed is known as the "half-life" and varies from a fraction of a second to hundreds of thousands of years.
Reprocessing	Separating the remaining fissionable fuel from the radioactive waste in spent fuel.
SABR	Subcritical advanced burner reactor.
SABrR	Subcritical advanced breeder reactor.
SMR	Small modular reactor.
Spent Nuclear Fuel (SNF)	Used nuclear fuel removed from a reactor to improve overall reactor performance before all the fissionable material has been fissioned.

Subcritical Reactor	A nuclear reactor in which more neutrons are being lost (by being captured or leaking out) than are being produced by fission.
Thermal Reactor	Nuclear fission reactors in which the fission neutrons are moderated from the million electronvolt energy range of fission neutrons to the less than 1 eV range where most of the reactions take place. Pressurized and boiling water reactors are the major thermal reactor types in the world today.
Tokamak	The leading fusion plasma magnetic confinement device.
TRU	Transuranics (isotopes heavier than uranium).
Transmutation	Conversion of atoms of one type into atoms of another type.
Tritium	An isotope of hydrogen with one proton and two neutrons in the nucleus.
Tritium Breeding Blankets	Tritium is radioactive with a 13-year half-life, so fusion reactors must make their own tritium by neutron capture in Li in "tritium breeding blankets" containing lithium placed around the plasma chamber.

References for "Fusion–Fission Hybrids"

[1] Global Energy Statistical Yearbook (2018).
[2] B. Richter, *"Beyond Smoke and Mirrors (Climate Change and Energy in the 21st Century)"*, 2nd ed., Cambridge University Press (2014).
[3] Nuclear News, p. 33 (March 2019) and p. 62 (March 2022).
[4] S. Gallier, Nuclear News (May 2020).
[5] W. M. Stacey, *"Nuclear Reactor Physics"*, 3rd ed., Wiley-VCH, Weinheim (2018).
[6] "Nuclear Fuel Cycle Overview", World Nuclear Assoc., www.world-nuclear.org/information-library/nuclear-fuelcycle
[7] "Uranium 2018", OECD, NEA, IAEA (2018).
[8] "Uranium 2014", OECD, NEA, IAEA (2014).
[9] L. Lidsky, "Fission–Fusion Systems: Hybrid, Symbiotic and Augean", *Nucl. Fus.* 15, 151 (1975).
[10] R. Moir, J. Lee, M. Coops, *et al.*, "Fusion Breeder Reactor Design Studies", *Nucl. Techn. Fusion*, 4, 589 (1983); also R. Moir, J. Fusion Energy, 2, 351 (1982).
[11] J. Källne, D. Ryutov, G. Gorini, C. Sozzi, and M. Tardocchi, "Fusion for Neutrons and Subcritical Nuclear Fission", AIP Intl. Conf. Proc., 1442 (2011).
[12] W. M. Stacey, "Fusion–Fission Hybrid Burner Reactor", p. 31, AIP Intl. Conf. Proc., 1442 (2011).
[13] R. W. Moir, N. N. Martovetsky, A. W. Molvik, D. Ryutov, and T. C. Simonen, "Mirror-Based Hybrids", p. 43, AIP Intl. Conf. Proc., 1442 (2011).
[14] W. Manheimer, "Fusion Breeding for Mid-Century Sustainable Power", *J. Fusion Energy* 3, 199 (2014). C. L. Stewart and W. M. Stacey, "The SABrR Concept for a Fission–Fusion Hybrid 238U to 239Pu Fissile Production Reactor", Nucl. Technol., 187, 1 (2014).
[15] "Accelerator-Driven Systems (ADS) and Fast Reactors (FR) in Advanced Nuclear Fuel Cycles", OECD/NEA, Paris (2002).
[16] M. Todosow, Idaho National Laboratory Report INL/EXT-14-31465 (2015); R. Wigeland, T. Taiwo, H. Ludewig, *et al.*, "Nuclear Fuel Cycle Evaluation and Screening – Final Report", FCRD-FCO-2014-000106, INL/EXT-14-31465, Idaho National Laboratory (2014).
[17] C. E. Till and Y. Chang, *"Plentiful Energy"*, Create Space, Charleston (2011).
[18] R. A. Knief, *"Nuclear Engineering"*, Taylor & Francis (1992).

[19]   J. R. Lamarsh and A. Barratta, *"Introduction to Nuclear Engineering"*, 3rd ed., Prentice-Hall, New Jersey (2001).

[20]   N. E. Todreas and M. S. Kazimi, *"Nuclear Systems I and II"*, Prentice-Hall, New Jersey (2005).

[21]   R. N. Hill, T. A. Taiwo, J. Stillman, *et al.*, "Multiple Tier Fuel Cycle Studies of Waste Transmutation", Proc. 10th Int. Conf. Nucl. Engr. (ICONE 10; 2002).

[22]   E. A. Hoffman and W. M. Stacey, "Comparative Fuel Cycle Analysis of Critical and Subcritical Fast Reactor Transmutation Systems", *Nucl. Techn.*, 144, 83 (2003).

[23]   W. M. Stacey, "Tokamak D-T Fusion Neutron Source Requirements for Closing the Nuclear Fuel Cycle", *Nucl. Fus.*, 47, 217 (2007).

[24]   C. M. Sommer, W. M. Stacey, B. Petrovic and C. L. Stewart, "Transmutation Fuel Cycle Analyses of the SABR Fission–Fusion Hybrid Burner Reactor for Transuranic and Minor Actinide Fuels", *Nucl. Techn.*, 182, 274 (2013).

[25]   W. M. Stacey, "Solving the Spent Nuclear Fuel Problem by Fissioning Transuranics in Subcritical Advanced Burner Reactors Driven by Tokamak Fusion Neutron Sources", *Nucl. Techn.*, 200, 15 (2017).

[26]   W. M. Stacey, C. L. Stewart, J. Floyd, *et al.*, "Resolution of Fission and Fusion Technology Integration Issues: An Upgraded Design Concept for the Subcritical Advanced Burner Reactor", *Nucl. Techn.*, 187, 15 (2014).

[27]   D. L. Jassby and J. A. Schmidt, "Electrical Energy Requirements for ATW and Fusion Neutrons", Princeton Rep. PPPL-3438, Hemisphere Publishing, New York (2000).

[28]   W. M. Stacey, "Capabilities of a DT Tokamak Fusion Neutron Source for Driving a Spent Nuclear Fuel Transmutation Reactor", *Nucl. Fusion,* 41, 135 (2001).

[29]   W. M. Stacey, *"Fusion: An Introduction to the Physics and Technology of Magnetic Confinement Fusion"*, 2nd ed., Wiley-VCH, Weinheim (2010).

[30]   W. M. Stacey, *"Fusion Plasma Physics"*, 2nd ed., Wiley-VCH, Weinheim (2012).

[31]   T. J. Dolan, *"Magnetic Fusion Technology"*, Springer, London (2013).

[32]   J. P. Freidberg, *"Plasma Physics and Fusion Energy"*, Cambridge University Press, Cambridge (2010).

[33]   R. J. Goldston and P. H. Rutherford, *"Introduction to Plasma Physics"*, Inst. Phys. Pub., Philadelphia (1995).

[34]   www.eurofusion.org

[35]   www.ITER.org

[36]   "International Tokamak Reactor": STI/PUB (#556, 1980; #638, 1983; #714, 1986; #795, 1988), IAEA, Vienna.

[37]   W. M. Stacey, "The INTOR Workshop – An Unique International Collaboration in Fusion", *Progr. Nucl. Energy*, 22, 119 (1988).

[38]   W. M. Stacey, *"The Quest for a Fusion Energy Reactor"*, Oxford Press, New York (2010).

[39]   www.aries.ucsd.edu.

[40] Y. Wu, "Fusion–Fission Hybrid Reactor Research in China", Conf. AIP 1442, Frascati ENEA (2016).

[41] V. Kuteev, "Development of Tokamak Based Fusion Neutron Sources and Fusion Fission Hybrid Systems", Conf. AIP 1442, Frascati ENEA (2016).

[42] Y. S. Shpanskiy and the DEMO-FNS Project Team, "Progress in the design of the DEMO-FNS hybrid facility", *Nucl. Fus.*, 59, 076014 (2019).

[43] W. M. Stacey, "Tokamak D-T Fusion Neutron Source Requirements for Closing the Nuclear Fuel Cycle", *Nucl. Fus.* 47, 217 (2007).

[44] W. M. Stacey, W. Van Rooijen, T. Bates, *et al.*, "A TRU-Zr Metal-Fuel Sodium-Cooled Fast Subcritical Advanced Burner Reactor", *Nucl. Techn.*, 162, 53 (2008).

[45] T. S. Sumner, W. M. Stacey and S. M. Ghiaasiaan, "Dynamic Safety Analysis of the SABR Subcritical Transmutation Reactor Concept", *Nucl. Techn.*, 171, 123 (2010).

[46] C. M. Sommer, W. M. Stacey and B. Petrovic, *Nucl. Techn.*, 172, 48 (2010).

[47] C. M. Sommer, W. M. Stacey, B. Petrovic and C. L. Stewart, "Transmutation Fuel Cycle Analyses of the SABR Fission–Fusion Hybrid Burner Reactor for Transuranic and Minor Actinide Fuels", *Nucl. Techn.*, 182, 274 (2013).

[48] C. L. Stewart and W. M. Stacey, "The SABrR Concept for a Fission-Fusion Hybrid U238-Pu239 Fissile Production Reactor", *Nucl. Techn.* 187, 1 (2014).

[49] A. T. Bopp and W. M. Stacey, "Dynamic Safety Analysis of a Subcritical Advanced Burner Reactor", *Nucl. Techn.*, 200, 250 (2017).

[50] "MATLAB", MathWorks website: https://www.mathworks.com/products/matlab.html

[51] www.statistica.com/statistics/lithium

[52] "Mineral Commodity Summaries 2019", US Geological Survey Report (2019).

Topical Summary

www.ingramcontent.com/pod-product-compliance
Lightning Source LLC
Chambersburg PA
CBHW050512190326
41458CB00005B/1509